MORALITY, DECISION AND SOCIAL ORGANIZATION

VIENNA CIRCLE COLLECTION

VOLUME 6

EDITOR: HENK L. MULDER

KARL MENGER

MORALITY, DECISION AND SOCIAL ORGANIZATION

Toward a Logic of Ethics

D. REIDEL PUBLISHING COMPANY

DORDRECHT-HOLLAND / BOSTON-U.S.A.

MORAL, WILLE UND WELTGESTALTUNG. GRUNDLEGUNG ZUR LOGIK DER SITTEN

First published by Julius Springer, Vienna, 1934

This edition is based on a translation by Eric van der Schalie

Library of Congress Catalog Card Number 74–81941

Cloth edition: ISBN 90 277 0318 3
Paperback edition: ISBN 90 277 0319 1

Published by D. Reidel Publishing Company,
P.O. Box 17, Dordrecht, Holland

Published in the U.S.A., Canada, and Mexico
by D. Reidel Publishing Company, Inc.
306 Dartmouth Street, Boston,
Mass. 02116, U.S.A.

Printed in The Netherlands by D. Reidel, Dordrecht

CONTENTS

CHAPTER I

COMMON TRAITS OF THE CLASSICAL SYSTEMS OF ETHICS

AN INTRODUCTORY LETTER ABOUT WHAT WILL NOT BE SAID

You are surprised – as I gather from your letter – to find me investigating problems of ethics. Actually these questions have occupied me for some time. That I have intensely, and in appearance completely, applied myself to mathematics all these years has, of course, been mainly due to the immense attraction of abstract ideas for their own sake. But not entirely. I have also been motivated by the thought that logico-mathematical studies are an indispensable training for most other branches of intellectual activity and, strange as it may seem to you, especially for the treatment of questions of morality. Indeed, the applicability of exact thinking to ethics appears to me to be an aspect of science that is of some importance for human life.

In addition to surprise, you express a two-fold apprehension: that, on the one hand, my temperament may misguide me to make rash subjective evaluations, even though ethics strives for objective knowledge; and that, on the other hand, my exact mode of thinking may lure me into a purely intellectual treatment, whereas morality aims at shaping behavior on the basis of the deepest feelings.

I have been aware of these dangers; and, in accordance with your warning, I will apply logic to ethical questions without allowing it to be influenced by subjective feelings. At the end of the book I shall make a brief assessment based on personal evaluations; but it will be marked as such and kept separate from theoretical considerations.

And the results of my involvement with ethical questions? You will come to know them from the following pages. In this letter I only want to call your attention to a few common traits of classical investigations into the subject that will *not* appear in my studies.

Above all, moral philosophers have believed that one of their main tasks is to search for "the concept of morality," to intuit "the essence of the good," to provide "the definition of obligations," to discover "the principle of virtue." I will deal with none of these questions.

Insofar as previous ethical systems have gone beyond analyses of concepts, intuition of essences, and the search for definitions, most of them have claimed to supply "foundations" of moral concepts and "justifications" of precepts. The creators of nonreligious doctrines based their maxims on "absolute values": the will of nature or the way to fulfilment; beauty or justice and duty; the best interests of men or the well-being of mankind, the nation, the state or the individual. In contrast, I propose to confine my cognitive studies to facts: to ascertain, to describe and to order them without saying anything about foundations or justifications; while in my value judgments – if you should wish to discuss evaluations at all – I shall rely exclusively and admittedly on my personal feelings.

No matter how much the normative attempts of philosophers, founders of great moral systems, and jurists may differ from one another, most of them have one trait in common: each system makes the claim to generality in the sense that what it demands is good and what it forbids is evil – good and evil for all men, universal in space and time. Human relationships vary, however, at least according to area and epoch – entire categories arise and pass away with technical progress – so that in other places and in later periods important aspects of life are not covered by earlier attempts. Now, for my part, I do not intend to present a particular moral system with a claim to universal validity. Neither do my thoughts lead me with cogency to any such doctrine, nor, in view of the actual conditions, is such a unified regulation in harmony with my feelings.

Even legal systems do not always quite keep up with changing relationships. Of course, in the interest of maintaining order, some newly emerging categories of human interconnections must be regulated by additional precepts, at least after they begin to appear with frequency. But it is precisely this limited elasticity of legal systems which causes many philosophers to neglect relationships controlled by law and to avoid a clarification of the relation between ethics and jurisprudence. It is a general characteristic of the spirit if not the letter of many ethical inquiries to assume that legal regulations deal with only the more gross social relationships, whereas moral ruling extends to the deeper relations and the innermost human problems. Thus higher morality is often said to begin with the judgment of what is legally indifferent, in other words, to start where the law ends; and the same spirit prevails with regard to political and economic systems and biological theories. Now, in practice, just the most significant human relationships are controlled by law and are connected with questions of politics, economics and biology. The neglect of legal and other practical subjects must, therefore, result in an ethics that does not treat those questions which stir modern man the most. I, on the contrary, shall proceed without any regard for the blurred lines between morality and the law, it being inconsequential to me either to obliterate them or to draw them clearly.

But do not expect from me material inquiries into politics, economics or biology. However important some topics in these fields are for ethics, my investigations are, as I have told you, concerned with the application of mathematical modes of thinking; and these deal only with the formal side of questions. The discovery and collection of empirical data, however decisive for practical tasks, fall completely outside of the limits that I had to set for the following studies.

Note finally that I take no interest whatever in the questions so extensively treated by most moral philosophers as to what ethics is, what is its object, and what are its tasks. Then, in all this, you have almost a compendium of the actual problems of conventional ethics with the assurance that I will not deal with any of those topics and will not take part in the endless discussions concerning them.

But at this point, I will conclude my letter and send you five short epistemological notes. They will not be assembled into a system. For much as I value systematic theories in mathematics and in mathematical sciences and eagerly as I have tried to develop such theories myself, I take exception to systematism in epistemology. Anyway, only a few very primitive epistemological remarks are needed in studying ethical questions. So I will rather loosely assemble those notes in a way that will give you a general idea about the contents of my ethical investigations – that is, if you still have the desire to hear about them after reading this long list of topics that I will *not* discuss.

CHAPTER II

FIVE EPISTEMOLOGICAL NOTES ABOUT GOOD AND EVIL

1. THE DEVELOPMENT OF A PERSON'S SENSE OF MORALITY

The development of an individual's sense of morality begins at an early age; as a child he hears adults characterize some human actions and omissions as good and others as evil – all this with overtones expressing pleasure or displeasure. While educators aim at inducing their own use of the words 'good' and 'evil' in the young their principal objective is to guide the child's actual decisions toward that which they call good and away from that which they call evil. By means of rewards and punishments, promises and threats, schemes and suggestions, they try to connect pleasure with what they designate as good, and displeasure with what they designate as evil, or at least to create the expectation of some pleasure or displeasure. Incidentally, inducing pleasure by suggestion is not confined to the ethical domain; there are numerous aesthetical and even physiological instances. But the educator's ethical aim is particularly hard to achieve since in many cases it is in direct conflict with the child's primary experiences of pleasure and displeasure. Thus, in general, educators have only partial success with the child in regard to the use of the words, the actual decisions, and the accord between speech and actions; although discrepancies between words and deeds – characterized as hypocrisy – usually are especially censured.

In addition to the use of the words 'good' and 'evil,' the adolescent learns about the system of penal laws of the state in which he lives. Punishments are provided for many, though by no means all, modes of behavior his educators call evil as well as for certain behavior they do not call evil. Discrepancies between what is legally forbidden and what is termed evil come about in various ways. In order to make penal rules applicable, only modes of behavior are prosecuted that can be discovered even without admission of the defendant; but not, for example, what educators call evil thoughts, wishes and intentions, however strongly they may condemn them. For the sake of relatively precise formulations, laws stipulate exact numbers never mentioned in ethics: an age of discretion, an age of consent, statutes of limitation. In order to reserve the legal machinery for cases of practical significance the laws specify minimum amounts below which theft and, in Europe, the concealment of found objects is not prosecuted. Examples of this kind could be multiplied.

Furthermore, the youth hears his fellow citizens use the words 'good' and 'evil.' Various people employ them in not altogether identical ways, so that one and the same human action may be called good, indifferent and evil by different persons. But where the overriding majority of people with whom an individual is in direct contact agree in judging an action, this "public opinion," which does not always coincide with legal precepts, also contributes to his attitude.

Recognizing all these circumstances or influenced by strong emotional experiences, many an adult alters the general direction of his decisions as well as his use of the words 'good' and 'evil' in the course of his life – suddenly or gradually, once or several times. The principal cognitive elements prompting such changes are those connected with the realization of a lack of uniformity and precision in the use of the terms 'good' and 'evil': different uses by various persons; divergence between the uses by individuals, in the public opinion, and in legal decisions about right and wrong; differences between public opinions in various sectors of society; differences between laws in various nations; changes of laws and a direct or indirect influence of individuals on those changes; the obvious arbitrariness of the specific limits set by the law that have been mentioned. Of further influence is the recognition of historical, political, economic and biological facts and theories and, most of all, insights into conditions for social coexistence. The principal emotional elements are experiences after the period of upbringing is concluded: mainly in the spheres of the drive for power and of sexuality. For example, an individual raised to have a liberal attitude in erotic matters may alter his views under the influence of repulsive experiences or as a result of liberties of others toward himself, while a person brought up by the rules of strict monogamy may modify his opinions under the influence of an irresistible attraction to a third person.

In view of the practical consequences attending the expressions of moral judgments according to their conformity or non-conformity to laws and public opinion, many persons allow what they say concerning good and evil to deviate more or less from the direction of their actual decisions. In fact, in the case for example of some demagogues, utterances about morality vary according to the situation and the people they address. Moreover, a person's decisions are not always in line with his own wishes about the general nature of his behavior – wishes that have come about as a result of his upbringing and the described conditions. Thus, in addition to a person's actual decisions and to his remarks about good and evil, a third factor emerges: his wishes about his decisions and actions. This third factor is called the voice of his conscience.

With regard to various moral norms, a person's involvement is of different degrees. The norms concerning theft are important to every individual since everyone is almost constantly in a position to steal or to be stolen from. The rules concerning deviant sexual behavior are of slight importance for the majority of people. Rules of court etiquette are a matter of indifference to the commoner.

When learning the use of the words 'good' and 'evil' the child also learns to understand 'shall' and 'shall not', 'ought' and 'ought not'. It becomes clear that the words 'you shall do this' express the educator's desire that the designated action be performed while its omission may be punished. Religious commandments beginning with 'thou shalt' are presented as God's wishes concerning human behavior, the violation of which results in punishment. The child further learns that one ought always to do what is good and not to do what is evil. From a purely formal point of view, this means that one of the uses of the words 'shall' and 'ought' is parallel to the use of the word 'good.' Finally, the youth becomes acquainted with the commandments of the state as governing what shall and shall not be done. Closer examination thus reveals that the use of 'shall' is burdened with equivocations: 'you shall' may express the speaker's wish; it may relay the wish of a third party; it may express agreement between that which shall be done and a system of norms – a system perhaps accepted as an obvious point of reference and not even expressly stated; or it may be used synonymously with 'good.' Accordingly, the youth learns to express the question "Shall I do this?" in an equivocal manner: at times to mean "Do you wish that I do this?" or "Does a certain third party wish that I do this?" or "Is this in accordance with a system of norms (possibly not expressly stated)?" or sometimes to mean "Is this good?"

In trying to transmit their use of the words 'good' and 'evil' and in helping to shape the character of decisions of the young, adults usually compromise between the way they were brought up and the way they have changed. Thus, in ethical matters ontogenetic development, particularly if it is parallel in many persons of one generation, becomes the basis of phylogenetic change.

Beyond this remark, the phylogenetic development of the words 'good' and 'evil' will not be investigated, however interesting the history of these words or their equivalents in other languages may be for various purposes. Even if it could be shown, for example, that 'good' originally meant the strong or characterized the Gothic, for the purposes of the present book this would be of no greater importance than is the derivation of the Russian word 'czar' and the German word 'Kaiser' from 'Caesar' for the theory of government.

In summarizing the preceding pages, it should be noted in the *first* place that a normal individual, when confronted with human actions and omissions, in numerous concrete situations answers "yes" or "no" to the questions "Is this good?", "Is this not good (evil or indifferent)?", "Is this evil?", "Is this not evil (good or indifferent)?". But not in *all* situations. In some cases, after thinking for a long time, the individual will answer "I cannot say." It is the lines between what a person calls indifferent, on the one hand, and what he calls good and evil, on the other, that are often rather unsharp. Indeed, almost everyone, in the course of his life, is confronted with situations in which he cannot even answer that question for decisions of his own. Thus an individual does not classify *all* human behavior into groups of good, indifferent and evil. But every rational adult speaker of English relegates at any given point in time the majority of modes of concrete human behavior to one of those three categories. And every rational French-speaking adult undertakes a corresponding tripartition with the words '*bon*,' '*indifférent*,' and '*mauvais*.'

Secondly, there are actions and omissions with regard to which almost all people answer the question "Is this good? indifferent? evil?" in the same way, and others about which they disagree. This can be illustrated by considering the act of killing or maiming a man under various circumstances. When committed for the purpose of robbery, the act is called evil by almost everyone; if performed in the course of a duel, it is called proper by some Europeans, indifferent or evil by others; and as an act of war, it is called good by many, indifferent by others, and evil by few.

If we consider the possibility mentioned above of differences between a man's inner judgments, his expressions concerning good and evil, and his actual deeds, then we see that each individual at any point in time undertakes not only one, but at least three tripartitions of the modes of human behavior: into what he feels is good, indifferent or evil; into what he characterizes by these words; and into modes of behavior that he regularly follows, that he sometimes follows and sometimes avoids, or that he regularly avoids. If in an investigation *one* tripartition is spoken of in order to simplify the analysis, then the differences between the various classifications are disregarded or the study is confined to individuals for whom the three tripartitions happen to be identical. But it must be recognized that the divergence between the tripartitions of an individual and the parallelism of such divergences in various individuals represent important factors in more precise analyses and are interesting topics of investigation in themselves.

2. The ideals

In the first note, it may be said, only that which is *termed* good and evil was discussed whereas moral systems decree what *is* good and what *is* evil; and ethics aims at ascertaining what *is* good and evil. Since one is in the habit of saying "We should do that which is good and not do that which is evil," the question "What behavior is good?" has also been expressed in the form "What ought we to do?".

One answer is: *good is what is in accord with the will of nature*. But what *is* in accord with the will of nature? Everywhere in nature we see that the strong are victorious; nature wants the supremacy and the thriving of the powerful and the annihilation and disappearance of the weak. Up to a certain point, it is clear what code for human behavior corresponds to this view. But one may say just as well: sympathy with the weak is a feeling that nature has implanted in man. Obviously a code complying with this interpretation is entirely contrary to the one mentioned before. Or another example: it is the will of nature that everyone cultivate his piece of land and lead a simple life, as Tolstoy preached. But one can also say: nature wants man to use his intelligence and inventiveness to develop technology and eventually to explore and inhabit other planets. The statement 'good is what is in accord with the will of nature' thus is compatible with the most varied ideas as to what it is that corresponds to the will of nature and it therefore receives content only after one has made specific stipulations about the will of nature. Once such stipulations have been made, however, that statement is superfluous.

A second answer is: *one should do what is just*. But given a concrete situation, what is just? To give each person what is due to him. But in a specific instance, what is due to a person? A system of rules is indispensable, if only to the effect that a man's due is what a certain person or group of persons designates as that man's due. Whatever it may be, the answer to the original question obviously lies in the specific system of rules, while the statement that the behavior according to those rules is just or good is without consequence. In absence of additional rules or in the process of establishing rules, one can say about the elusiveness of the word 'just' all that has been said about 'good' in the preceding note. The reduction of 'good' to 'just' does not constitute any progress.

Other answers are the following: *good is the striving for perfection and fulfilment*; *good is the beautiful*. But what is perfection? fulfilment? the beautiful? The answer that is needed must not be in general terms of other ideals but must directly or by clear implication specify which decisions the pursuit of perfection demands in concrete situations. But as soon as a system of such rules has been specified it is again that system which is solely of significance and not the ideals on which it is based.

Yet another answer is: *one should do what duty requires*. This example is different from those previously mentioned, since Kant in his *Foundations of the Metaphysics of Morality* made the important attempt to establish an intimate connection between duty and one single formal principle. This principle is the so-called *categorical imperative*: "Always act according to that maxim of which you can wish that it become a general law." This book will not consider the epistemological nature or the reasons for the claim to validity of this principle or the other problems connected with the imperative that were important to Kant himself and to most other philosophers. Here, all that is relevant is the following question: what concrete precepts result from the categorical imperative in specific situations?

In an examination of this question, one factor above all becomes apparent: in order to be in a position to wish that a maxim become a general law one must first have a survey of the consequences which that law would entail. Thus before being able to apply the categorical imperative to a decision a person must engage in a cognitive activity. Even in rather simple everyday situations an approximate survey of the consequences of a particular rule – an exact survey is completely unattainable – can be achieved, if at all, only with great difficulty – difficulty to which a person of average intelligence is not equal. These difficulties are already almost insurmountable in the, as it were, static case of all men fulfilling their duty (obeying the categorical imperative); they become immeasurable in the dynamic case where there is a question of how a person should react to another individual's intentional disregard of his duty and conscious disobedience of the categorical imperative, or where the problem is to change laws and forms of social coexistence.

One result of these considerations, however, is clear from the outset: in almost every concrete situation there are several mutually incompatible decisions to consider, each of which, if it were made by all persons in the same situation, would lead to a possible coexistence of men. For instance, if a man asks whether he should be strictly monogamous, the categorical imperative can answer this affirmatively if the number of men does not exceed the number of women, for in this case it is completely practicable that all men live in strict monogamy. If he asks whether he might practice polygamy, then the categorical imperative cannot deny this if the number of women exceeds the number of men, since polygamy is known to have been a feature of great and powerful cultures, and thus certainly is a generally practicable social institution.

Actually, however, the categorical imperative, as stated above, does not merely direct one to act in such a way that his maxims could become general laws. This latter precept, according to Kant's *Foundations of the Metaphysics of Morality*, covers only the more immediate, strict, inflexible duties (although in his earlier *Critique of Practical Reason* he had called it a basic law). Rather, the categorical imperative commands one always to act according to such maxims as one *can wish* to be general laws. The above mentioned survey of the consequences of a general law is, therefore, only one prerequisite for the application of the imperative to a concrete case.

Imagine an individual having attained the necessary knowledge of the consequences of the various conceivable regulations of the situation. Which one can he wish to materialize as a general law? First of all, if 'can' here means logical possibility, then it includes all not self-contradictory laws, of which there exist in general more than one (although Kant does not seem to have realized this fact) and no specific decision is prescribed. Secondly, if 'can wish' means that which conforms to some sort of ideal, then, so that precepts may result from the imperative, those ideals must be named and defined (which Kant never did explicitly); and thereafter it will be essentially from that stipulation that concrete precepts will be derived. In the third place, if it is simply meant that the individual should act according to those maxims which he *actually* wishes to be general laws, then the derivation of concrete precepts from the categorical imperative would for each individual simply yield the laws that he himself desires.

As an extreme example, let a person apply Kant's principle to the question as to whether he should live in complete sexual abstinence. The categorical imperative states that he should live according to this maxim if he can wish continence as a general law. First of all, as to the pure logical possibility, it is clear that the general law would not be contradictory (as would be, for example, the logically impossible norm that all men should live in abstinence and all women should have heterosexual relations). Second, as far as ideals are concerned, the general law must be rejected by everyone who wants the human race to continue. On the other hand, for the misanthrope or extreme pessimist whose ideal is the extinction of humanity, suicide-through-continence would be one of the possible realizations. Third, whether or not someone actually desires that general law with all its consequences is a question that only he himself can answer and one which the overwhelming majority of mankind will answer negatively.

It is clear that in numerous cases that are not extreme and are, therefore, more meaningful in a practical sense, a plurality of rules can be devised each of which is free of contradiction, conforms to some ideals and is contrary to others, and, finally, is actually desirable to some persons and undesirable to others.

The preceding discussion must be supplemented by an important remark. If, given a choice, a normal person conforms his decision to what he would wish to become a general law (and hence identical to everyone's decision in the given situation), then he certainly will do so with the (perhaps tacit) proviso that the same decision will not necessarily be open to an imbecile, a public menace, or an invalid. He can even base this proviso on the categorical imperative by saying that a public menace placed in his situation would not represent the same case. An identical argument, however, may also be advanced by an individual who sees himself as exceptionally strong or otherwise unusually qualified or who belongs to some special group. He, too, may refer to the categorical imperative when allowing special maxims for himself or for equally qualified individuals or for members of his group. He may say, "I act according to that maxim which I wish to become general law, that is, to serve as regulation in all cases identical with mine. If, however, another person is confronted with my present situation, then an identical case obtains only if that person is as strong as I am or equally qualified, or a fellow member of my special group." To be sure, another person can deny the first person's special qualifications or contest that those qualifications should be especially accounted for in the law in question. But at any rate it is clear that the categorical imperative yields concrete conclusions only in conjunction with stipulations as to which individuals, if any, should be given different consideration in which laws.

In summary, we see that without supplementary stipulations even the categorical inperative and the concept of duty fail to lead to concrete precepts, let alone to a unique system of precepts. It is further clear that if one specifies additional norms – be it in accord with general ideals or the will of special individuals or concerning unlike treatment of unlike types of individuals – then it is those added norms that are decisive and not the fact that one designates their consequences by the term 'duty.'

An answer to the question "What ought we to do?" that appears at first glance to be of a different type presupposes that a group of people has been specified – the nation or a smaller group or mankind as a whole – and then proposes that members should always act *for the well-being* or *in the interest* of that group. But where does the interest of the nation lie? In peace or in conquest and aggrandizement? In a uniform distribution of moderate wealth or in raising the total wealth at the price of inequalities? In uniformity or in the creation of an elite? And in the latter case, should physical or intellectual qualities or recognized merit characterize the members of the elite? The opinions of various people as to what constitutes the interests of a given group and of mankind as a whole diverge widely. In fact, doubts also beset the individual following the norm that everyone should act in his own interest. Even if one identifies a man's well-being with the fulfilment of his wishes at any moment (despite their conflict with some wishes of others), difficulties arise because some of his wishes are incompatible with one another and their fulfilment may fail eventually to produce what he calls his own well-being. Thus the fundamental problems concerning the word 'interest' are the same as those raised by the word 'good' itself; and to base the latter on the former is futile.

We thus see that nature, justice, perfection, as well as duty and the interests of any group of people are unsuitable as a basis for the definition of good and evil or as a guide in answering the question "What ought we to do?" as long as it is not stated precisely how these ideals should function in practice. Once this is done, however, then all that is relevant are those very stipulations. Their designation as ideals is nothing but a complicating verbal adornment.

As an illustration imagine the following: an isolated, enormous multitude, tired of conventional moral conditions and worn-out from the eternal strife resulting from their conflicting wishes, convene in a mass meeting, each with the purest intentions prepared even to forgo privileges and property. All are inspired solely by the desire to realize the good and the moral. They pass a resolution to make the well-being of the assembled group the guiding ideal in the reconstruction of their coexistence. But after a short time they will begin to argue over what constitutes the well-being of the group. If, after reminding themselves of their noble intentions, they decide on a *just* re-ordering of their relationships, then this will lead to a new, no less serious dispute.

3. The Logical Role of the Ideals

From a purely logical point of view, the process is as follows: Someone asks "What is good?". He is answered by the assertion of one of the ideals mentioned above, such as "Good is what is just." If he asks "What constitutes that ideal?", for example, "What is just?", then the other can possibly postpone the answer even further by inserting more words. In the end, however, if he wants to answer at all, he will have to answer with a concrete system of norms. The entire process thus amounts to calling the general content of those norms by the name of the ideal, say "just," and then declaring that what corresponds to that ideal is good. Accordingly, the ultimate answer to the question "What is good?" is "That which corresponds to such and such norms," even though the answer is often given with the logically unimportant interpolation of one or several ideals. That interpolation is pointless since it consists in the stipulation that two words, for example, 'good' and 'just,' should be used synonymously. But the logically minded person undertakes the definitory introduction of terms only for the purpose of shortening statements; in other words, so that the introduced terms may replace longer expressions, but certainly not merely to complicate the language with synonyms for terms that are short themselves.

If the question "What is good?" is answered by "Good is that which is worthy of love or that which one ought to love," then the role of 'worthy' is analogous to that of the ideals. In order to be more specific, one obviously must ask further "What is worthy of love? What ought one rightly to love?" If one is told "That one knows through intuition" or "That is evident," then, from a logical point of view the state of affairs can be described as follows: the question "What is good?" is answered, in the final analysis, with "This is evident," after the interpolation of the words 'worthy' and 'rightly.'

The following dialog is completely analogous in its logical structure:

Q. Which medicines are beneficial?

A. Those medicines are beneficial which are worthy of ingestion or, in other words, those which are appropriately used.

Q. Which medicines are worthy of ingestion? Which are appropriately used?

A. One learns that through experience.

In this conversation, the answer to the question "What medicines are beneficial?" (after the insertion of the words 'worthy of ingestion' or 'appropriate') clearly is: "This is learned through experience."

Of course, one could just as well turn things around. If someone asked "Which medicines are worthy of ingestion?" and received the answer "The beneficial ones," he might well ask further "Which medicines are beneficial?" and be told "That is learned through experience." But even this twist has an analogy in ethical terms.

Q. What is worthy of love?

A. That which is good.

Q. What is good?

A. That is evident.

Such discussions not only exist in the ethical literature but are considered to be important achievements of well-known moral philosophers.

The creator of a moral philosophy who agrees with the view set forth in this note will construct a system of norms that is as concrete as possible. The belief that concrete precepts for human behavior can be deduced or otherwise derived from general norms is almost always erroneous. It is probably due to insufficient insight into the logical relations between propositions and into the nature of proofs.

Another logical error lies in the belief that concrete norms can be proved indirectly by a demonstration that the assumption of their non-validity leads to contradictions. Strictly logical contradictions, very scarce in the realm of norms for human behavior altogether, are confined to trivial instances. For example, the rules that an object lost through theft remains legally the property of the loser, and that an object acquired by theft becomes the property of the state are contradictory in view of the value-free proposition that the objects lost through theft and the objects acquired by theft are the same. But again considerably more seems to be before the minds of those who believe that they can establish indirect proofs for moral norms; and again their claim is apparently based on an insufficient insight into logic. For in a strictly logical sense, in particular without appeal to value judgments of some kind, even such a primitive rule as "Thou shalt not murder" is incapable of proof.

The moral philosopher would be further confirmed in his quest for very concrete norms if he would ask which norms have actually been observed in the course of history. It has been the commandments of worldly and religious authorities, while those considered or formulated by moral philosophers have been of little direct influence. The reason for the prevalence of the former is not merely the threat that their violation will be punished before or after death, but also the concreteness of the authoritative precepts. Codes decreed by states also include rules for nonbelievers; religious systems contain precepts even for times of change of the civil authority or its collapse. Imagine, in contrast, the confusion in the legal and moral life of a community if it were to follow certain philosophers in establishing the following norm: "Man should be good. Good is that which is worthy of love. What is worthy of love is revealed by intuition."

Even the threat of the most terrible penalties for the violation of this commandment would not alleviate the situation unless, of course, the good were defined by means of additional concrete precepts, or unless the intuitive experience of that which is worthy of love were described or exemplified by citing concrete norms. However, as soon as this would happen, the actual regulation would obviously be based on those additional precepts, while the philosopher's contribution to this system of norms could be eliminated without any risk of being missed. And, as discussed in the second of these five notes, the situation would be essentially the same in a state which set up the commandment: "Every person shall do his duty. Duty is acting according to the categorical imperative."

Or imagine a community which, following Nietzsche, has decreed a double standard by the norm that every master is allowed to satisfy all his wishes while the others must serve the masters. It is clear that without the establishment of exact criteria as to whether or not any given member of the community is a master, the administration of justice in that community, too, would be impossible, and moral life would find itself in a state of utter confusion. What Nietzsche himself seems to have had in mind was the criterion that a master (*Herrenmensch*) is he who considers himself as such, that is, he who assumes the prerogatives and puts himself above slave morality. Pursuing, on the basis of this criterion, the consequences of Nietzsche's norm one sees that masters could easily codify a morality for the slaves; but one is at a loss as to how to settle disputes between masters, each of whom acts as he pleases. It is clear that the coexistence of masters, if there be more than one, presupposes a substantial limitation of their freedom. Yet, one can imagine a situation, where, in addition to a system of laws for slaves, a looser code for masters is established which only restricts their behavior in relation to one another. In fact, such a system was actually in force in antiquity and, to some extent, in the Middle Ages.

But now assume that a master is the victim of an act such as an insult that is a violation of the code when committed by a slave. When apprehended, the culprit, whoever he may be, can avoid punishment, or at least punishment by slave laws, just by asserting that he himself is a master. As proof of this claim he could, according to Nietzsche's criterion, simply quote his courage in committing the deed in question. Thus that criterion would make the application of slave laws illusory; or, in other words, from a purely logical point of view, Nietzsche's criterion is incompatible with his norm. To put it in still another way: if double morality should be practicable, then one must formulate criteria independent of the judgment of the persons involved as to which set of laws is applicable to whom; for example, precise external characteristics such as parentage, physical or mental qualities, or recognized merits. Nietzsche's criterion, in contrast, is an unsuitable basis for the regulation of moral and legal life. This, incidentally had already been recognized by Plato, as can be seen from the answers that he let Socrates give to Callicles in the Gorgias dialog.

The need for making norms as concrete as possible is still further underlined by the observation that even so-called concrete norms allow considerable room for arbitrary judgments about application and interpretation, since they include generic terms with more or less blurred limits.

In this connection, a few words should be devoted to those moral philosophies which emphasize a *virtue* such as charity or fairness, rather than a general *ideal* such as the interest of mankind or justice. The commandment to practice a particular virtue entails more concrete precepts than the norm to follow a general ideal – at least in a certain group of situations. Charity commands, for example, to someone sitting at a full table that he give food to the hungry onlooker; to the idle that he help an overburdened man to carry his load; to everyone, that he support the ill and the weak. Notwithstanding their simplicity, such elementary rules cannot be inferred from the general ideal to act for the good of mankind except in conjunction with specific stipulations as to what constitutes the good of mankind; while in conjunction with contrary stipulations the general precept may well imply that one should hasten the elimination of the weak and the ill. Yet even the commandment to practice a virtue fails to supply rules for behavior in *all* situations or *unique* precepts for decisions in all cases to which it is applicable. A man with a large fortune who decides to practice the virtue of charity cannot deduce the economic acts implementing his decision: whether to give all his belongings to the poor or to manage his fortune profitably and give away his profits, to mention only two possibilities. And does the virtue of charity prescribe that a person yield to the passionate desires of another or respect the pangs of jealousy of a third person? Or what other behavior is he to adopt?

Obviously, there are numerous important dilemmas where the precept to practice even a relevant virtue does not provide a decision.

This state of affairs is of course understandable through an examination of the epistemological role of commandments to practice a virtue. Each one of such words as 'charitable' or 'compassionate' or 'fair,' serves to characterize modes of behavior in certain situations, but definitely not modes of behavior in all situations, nor unique modes of behavior in all situations to which the word applies. In some cases it describes mutually incompatible modes of behavior.

Applicability of the word 'charity' to all or even only to all practically important situations (for example, to questions connected with justice or veracity) is, of course, out of the question. Nor is the word used in a strictly defined or unique sense where charity is relevant. All this lies beyond the scope of everyday language. Virtue ethics do not include precise definitions of their respective central virtues by systems of norms; and if they did it would be those norms that would be the point of departure for judgments, while the fact that fulfilment of the norms is designated by the name of the virtue would be insignificant. Unless the concept of a specific virtue is precisely defined, however, all that the norm to practice the virtue prescribes is the performance in those cases to which the virtue is applicable of one of the actions compatible with the use of the name of that virtue. Clearly in many important situations no specific action is prescribed, and in many other situations several decisions are possible. Even, so, the commandment to practice a specific virtue is much more concrete than the norm to follow a general ideal.

4. The Essence of the Good. The Meaningless

"Still nothing has been discussed but what it is that people *call* good or evil," it will be said. "But what *is* good or evil? Is not a deed or an omission good or evil or indifferent apart from and independent of whether it is *termed* good or evil or indifferent? Even if there should not exist an idea of good residing somewhere in a Platonic universe so that the goodness of an act is determined by its share in that idea – at least there must pre-exist the species of good deeds and omissions so that we can combine those modes of behavior into a group and call each of them good."

Whether anything *is* good apart from and independent of what is *called* good is a question that we shall answer neither affirmatively nor negatively; it is a question that we shall not discuss. We confine ourselves to the statement of the facts regarding the actual use of the words 'good' and 'evil' by the maturing person as described in the first note; and we will say about good and evil nothing beyond that.

In particular, it is far from our aim to say "*Good is what is called good.*" It would, therefore, be a complete misunderstanding of our view to criticize it by pointing out its incompatibility with the fact that there are good deeds which are not called good – deeds which in fact are performed so discreetly that they never come to be judged at all; just as there are also deeds that may not be good but are called good by some (evil) persons. In the *first* place, we would never say "Good is what is called good" because of the lack of precision of that sentence. Its formulation does not make clear whether it is meant that good is an act that is termed good by one person or by several people or by everyone; and these are three quite different groups of actions. But neither do we intend to assert any of the three corresponding particular sentences. For each of these sentences is supposed to determine the use of the words 'good' and in doing so makes reference to what is called 'good.' We thus consider them – and this is our *second* reason for eschewing those sentences – either to be circular or to introduce an equivocal use of 'good.' While the introduction of synonyms is contrary only to the aesthetical and economic intentions of the logically minded, there are still much more important arguments against the introduction of equivocations. In the *third* place, we would shun such sentences even if they were construed as introducing, besides the word 'good,' the use of something like a higher level good; for example, by stipulating that the *truly good* is what the majority of men call good and only that. We don't feel the need for the definitory introduction of good on a higher level. For the same reasons (and, if possible, even more strongly) do we reject the motto of classical relativism: "*Good is for each individual what he regards as good.*" All that we did establish in the preceding notes was the simple fact that certain modes of behavior are called good or evil; we investigated how the use of these terms develops in an individual; we pointed to differences in the way the words are used by various individuals and by one individual at various times; and we have limited ourselves to these facts.

"But what is the *essence* of good and evil?" someone may ask. "True, this cannot be established through experience, but certainly it reveals itself to our *intuition* (or what Husserl calls *Wesensschau*)."

"What is loved in *correct* love? This question does not aim at finding out what, according to our experience, is actually loved; rather, we are asking what is *worthy* of love. This is quite independent of what is *termed* good; but it is *evident*."

"What is valid in the domain of *duty*, in the realm of *values* with its hierarchy? This has nothing to do with wishes and desires, but can be recognized by cognitive *insight*."

"The sentence 'this is good' differs in *significance* from an expression of feelings; it means something different from 'this deed pleases me' and 'this deed pleases many people' and 'this deed pleases everyone.' It means something objective, just as does the sentence 'this lawn is green'; and we recognize *what* it means through a special sense."

Anyone making such assertions is irrefutable insofar as he is speaking of his subjective experience; and we will not attempt to argue with him or to deny the possibility that his assertions are true in some sense. We merely claim for ourselves the right granted to everyone, to be the final authority in the description of his own experiences. We have feelings and desires concerning certain modes of behavior; and we can express them using the words 'good' and 'evil.' But it is not in the last analysis evident what is "worthy" of love. We have no cognitive insight into the "essence" of good and evil beyond this, nor an intuition into a realm of values with a definite hierarchy, or a special sense for what is meant by value judgments beyond and independent of the conditions described. Therefore, we will not discuss assertions based on these faculties with anyone who claims to possess them.

Are then these questions, which we have put aside, and the answers that others have given to these questions meaningless? This question, too, we refuse to answer either affirmatively or negatively; we will simply not discuss it. For woe to him who declares statements to be meaningless! He resembles a man who notices his mortal enemy inescapably mired in a swamp and instead of going his way, rushes toward him with drawn dagger.

It is obvious without further comment that what has been said concerning the good can be expanded to statements about value judgments in general.

Even beyond that, however, these epistemological considerations extend into questions concerning the foundations of mathematics and logic. Some of these questions have been investigated by the author elsewhere [in the late 1920's] along lines diverging from all other tendencies then prevailing in studies of the foundations of mathematics and logic. Statements of the form "Such and such methods, procedures, modes of reasoning, axioms and concepts are meaningful, while such and such others are meaningless" have recently [in the early 1930's] become quite common. The author has regarded statements of that form as value judgments and as such relegated them to the biographies of their proponents while suggesting that the expression 'mathematical theory' be used for the transformation – free from praise and censure, license and proscriptiom – of certain propositions into other propositions with the help of precisely stated rules.

5. The Development of the Epistemology of Morality

The epistemological view presented in the previous discussion with regard to an ideal (such as justice), an objective value, or an absolute good develops in five stages.

In the first stage, the uncritical person designates as justice a system of norms that he has inherited or, in some cases, a code that he has modified or even norms that he has created.

In the second stage, he asks "*Is* this justice?" and then "*What is just*?".

In the third stage, he recognizes that various answers to the latter question are actually available and he asks "How shall I recognize, or who shall judge, what is just?"

In the fourth stage, he sees that the matter is not of cognitive nature but rather a question of a decision; and he asks "How shall it be established, or who shall establish what is just?"

In the fifth stage, he accepts a system of norms or seeks to modify a code or even to replace it with a new one. With greater or lesser intensity he *desires* certain modes of behavior and conditions in his surroundings or within a larger group of people or, possibly, within mankind; and he *acts* to realize these desires with more or less energy – at any rate, however, without speaking of justice at all.

CHAPTER III

THE ETHICS OF DECISIONS

A DIALOG ON DEMYSTIFIED ETHICS

1. *Whether investigations according to the principles suggested in the preceding notes belong to ethics at all.*
2. *Whether there do not exist still other ethical questions.*
3. *Whether ethics is analogous to geometry.*
4. *Whether systems of norms might not be combined by logical operations.*
5. *Whether decisions are the only basis for morality.*
6. *Whether rational foundations for decisions are possible.*
7. *What role faith plays in morality.*
8. *What demystified ethics might be able to achieve…*
9. *… except for a logic of norms,*
10. *… and except for a logic of desires.*

(1) "Having read your letter and the five notes you sent me I admit that you have secured one point: an ethical system developed according to your principles will not include a determination of what is good that anyone might find false, no matter what his point of view. But what price have you paid? Along your way, *nothing at all* has been declared good; what good is has not been defined; it has not been stated what the essence of good is; or, to put it in yet another way, the good is simply not being discussed. The price, then, is that the ethics which can result from your principles is an ethics without good and evil, without morality and immorality. But is this not also a judgment of your work? Are yours not ethical aspirations that have missed their mark? Can one even call such a study *ethics*? Is it not comparable to a zoology that does not deal with animals, an astronomy that ignores stars?"

My dear friend, I do not value delimiting any science. You will never hear me talk about the existence or nonexistence of the "ideal subject of each individual science"; about "the objective organization of the realm of truth into compartments, the spheres of the several sciences;" about "the points of departure of the individual sciences, to be grasped through insight into their essences." Much as has been said about these and related problems of methodology – for my part, I avoid them as I would avoid the swamp that I mentioned in a simile. I will not speak of these questions either to assert or to disavow. In my opinion, the name of a science (and more generally, of a *Wissenschaft*) is a word that is useful and indeed at times indispensable, as a brief designation of a certain group of propositions – but of a group changing with time and of propositions somewhat loosely connected by the terms in which they are formulated as well as by all sorts of mere historical accidents. For example, a new topic is often classified as belonging to a certain science because it has been brought up by persons who have previously dealt with topics already assigned to that science. Consider, for instance, geometry which originally dealt with measuring the earth, as its Greek name indicates. Even in its early stages, propositions were incorporated into geometry that have no more to do with measuring the earth than with, say, organizing a system of messengers. Later, 'geometry' came to be used for the system of quantitative statements concerning space. Still later, however, geometers (which at that time meant men who were familiar with and investigating quantitative statements about space) discovered propositions about space that had nothing to do with quantity, such as statements about the arrangement of points on straight lines and circles or about the differences between a line, a surface, and a solid. These propositions were also assigned to geometry. But a precise definition or even a uniform use of the word 'geometry' has never been achieved. Geometry certainly has not fared the worse for that fact. Delimitations of geometry with regard to subject and method would not only be impracticable but also completely useless because some of the most interesting new discoveries of geometers would always break through the limitations. Since I cannot name an ideal object even for geometry; since I regard a delimitation as impracticable and pointless; and since the use of the word 'geometry' is subject to unending modifications, it will not surprise you that I adhere to this same point of view all the more strongly with regard to ethics. Whether or not my investigations that are free of the words 'good' and 'evil' are assigned to ethics is a matter of no particular concern to me. Yet I cannot see why they should, as a matter of principle, be excluded. They certainly include terms that are characteristic of statements of classical ethics. But if you nonetheless insist on excluding them, I shall not try to prevent you from doing so; for, as I have said, I view the entire question, which only leads to a terminological quarrel over the use of the word 'ethics,' as completely unimportant, and I do not wish to pursue it further. You can be assured, however, that I shall by no means be as inconsistent as are those who forget the principles they have applied to a particular science when they discuss science as a whole. Those who speak of the unity of science (and in German have even created the monstrous word '*Einheitswissenschaft*') seem to overlook the fact that the separation of science from other intellectual activities can be no more practicable or useful than that of particular sciences from one another.

(2) "Fine. Let us pass by the question of whether or not reflections according to your principles should be included in ethics. What I wish to stress quite emphatically, however, is a point that is by no means merely terminological. Not only do you fail to study what *is* good and evil; you do not even consider the question as to what one *calls* good and evil. The statement *that* certain modes of behavior are termed good and certain others are termed evil, and the description of how an individual comes to use these words and how various persons deviate from one another in using them – all this cannot be considered as a treatment of the question, *what* is termed good and evil. The same method that you employ in discussing 'good' and 'evil' you could also employ, for example, in examining the word 'dog.' You can establish that some beings are called dogs; you can describe how a person comes to use the word 'dog'; and you can observe that various persons concur in using it or, perhaps, that in some border line cases they differ. The zoologist, for example, calls certain animals dogs that the layman would not, and does not designate certain other animals as dogs whereas the layman would. In order not to become involved again with questions of terminology, let us leave undecided whether such matters even belong to the realm of zoology – in my opinion they definitely do not. One point, however, remains incontestable: in no way does zoology consist *only* of such statements and descriptions."

Of course!

"Zoologists assume it to be universally known that certain beings are called dogs. One of their tasks, however, zoologists agree is the *description* of dogs – the citing of as many characteristics as possible of the animals so called. And accomplishing this task is not at all a matter of terminology or classification, but rather a matter of establishing facts: particular characteristics often coexist in one animal and will be found in its offspring, too. These are facts that are of course entirely independent of the names of the beings in which the attributes are found. One may add the linguistic fact that in English the animals exhibiting those characteristics are called dogs. But then where is your description of what is called good and evil? I cannot conceal from you that as a result of your not discussing this question your investigations seem to me to be superficial. Why don't you, for example, discuss ethical hedonism? Naturally I do not mean that kind of hedonism which merely states "Good is what results in pleasure," or which exhausts itself in variations and imagined proofs of this proposition; for your discussion of ideals has convinced me that this is all empty verbiage. Rather, I mean those hedonistically colored investigations that offer a few empirical rules as to which modes of behavior will, under certain circumstances, result in pleasure or happiness. For it is a fact that the realization of some wishes results in a lasting feeling of satisfaction and their nonrealization creates permanent dissatisfaction; while the realization of other wishes may bring momentary delight, but is followed by dissatisfaction and feelings of regret. These are facts about which one can make statements of the same degree of validity as about any other psychological subjects. But in only one passage of your notes do you touch – barely touch – on this question of inner contentment by mentioning the voice of conscience. This is a point on which I should think you could make more precise empirical statements and I miss them when you ignore the psychology and biology of morals. Or do you think that the behaviors called good have no common characteristics at all, and that, therefore, no factual statements on the subject are possible? By limiting yourself to a description of how the use of the words 'good' and 'evil' comes about, do you mean to suggest that the whole problem is nothing more than a matter of words?"

I value every contribution of historical and ethnographic facts to the study of what has been termed good and evil at various times and by various peoples. Especially do I ascribe the greatest importance to all that which you characterize as the psychology and biology of morals. Investigations in these areas establish empirical facts and regularities. More specifically they concern the interrelation between behavior of individuals and groups, on the one hand, and man's organism, his mental and emotional condition, his achievements, especially cultural achievements, as well as hereditary factors, on the other. Not only are many such facts known, some since antiquity, but I do not doubt that an even greater number will be revealed through further investigations. If I pass over these questions, then it is only to leave their examination to more qualified investigators. Remember that I do not intend to write a systematic book on ethics, but rather something much more specialized: as I have written to you, I want to apply mathematical thinking to problems of ethics. Accordingly, I address myself primarily to the formal side of ethical questions.

(3) "You probably consider the truths of morality and ethics to be intelligible in the same way as those of *geometry*. This was clearly Plato's thought when, in a dialogue concerning virtue, he turned the conversation to geometry, suggesting that man somehow evokes its truths directly from within himself."

It is not my belief that there is a far-reaching analogy between geometry and ethics, between space and morality. But I think that knowledge of geometry can be of use to the moral philosopher. Yet I am afraid I shall disappoint you. For just that which you seem to expect, geometry cannot provide: logic and mathematics cannot express preferences for certain systems of norms and still less designate a single system as "the true one." That entire expectation rests on a notion of geometry that does not correspond to our present views. Indeed it was refuted more than a century ago.

"You must not assume that I, a nonmathematician, have any special knowledge of geometry. Please explain, therefore, your last remarks in somewhat more detail. Am I to understand that you doubt the truths of geometry?"

One cannot speak of "the geometry" as did Kant and even the geometers of his time. They were thinking of the system of propositions that Euclid had deduced from a few axioms in his celebrated book, *The Elements*. Those who speak of one geometry today are unaware of the fact that soon after Kant, at the beginning of the 19th century, other systems of propositions were deduced from other geometric axioms, one of which contradicts an axiom of Euclid. Later, by modifying several Euclidean assumptions, geometers developed numerous "non-Euclidean" geometries. Some of their basic assumptions, and hence many of their theorems, contradict Euclidean statements; yet each of these geometries in itself is a consistent system of propositions.

"I am not sure that I understand you correctly. I am aware of the fact that geometric axioms are not proved. In fact, one *defines* an axiom as a proposition that *cannot* be proved, and, as one quickly adds, that *need not* be proved. But a geometric axiom is not in need of a proof because its truth is self-evident. So how can the opposite of an axiom be the point of departure for a geometric theory?"

The thesis that axioms are propositions that neither can be nor need to be proved was generally accepted for centuries and can still be found in some textbooks. But it originates in an interpretation of deductive systems that is completely obsolete. Since the regressive process of proving propositions must break off somewhere, every deductive theory begins with several unproven propositions which constitute the starting point of the deductions. They are called the axioms or postulates of the theory. This is what is true in your statement that axioms cannot be proved. As to the alleged validity and self-evidence of axioms, let me say this. One selects as axioms about undefined lines, for example, propositions asserting relationships that have been observed between lines drawn on paper, between thin, rigid rods, and between light rays. Geometric deductions merely state: *if* such and such relations obtain between undefined lines or lines on paper or rods or light rays, *then* also such and such other relations obtain. And these implications, which represent the sole content of geometry, are independent of experience. Nor, for reasons which would lead us too far from the main topic of our conversation, does our selection have anything to do with the fact that Euclid's axioms have become what is called self-evident. It is also quite possible to choose other axioms and to develop conclusions from *them* in a purely deductive manner.

"But is one justified in referring to deductive systems that proceed from other axioms as *geometries*?"

This is a terminological question which the actual linguistic usage has settled in the sense that I have described in discussing the use of the names of sciences and especially of the word 'geometry.' The first non-Euclidean geometry contained the same terms as Euclid's *Elements* (if partly in a somewhat different meaning) and was developed by men known to be geometers. To consider this system of propositions as a separate science would have been impractical, all the more since numerous non-Euclidean theories soon appeared. And while the newer theories differed increasingly from Euclid's geometry each of them was still connected with the earlier ones in two ways: by the terms in the propositions – notice, I say merely *terms*, for the *meanings* of the terms had been modified more and more (though of course in a precise manner!) – and by the persons creating the new theories. This is why they have always been referred to as *geometries*. More recent non-Euclidean geometries differ from Euclid's so extensively that except for their verbal and personal or historical connections there really is no reason for calling them geometries; and indeed the limits for the use of the word 'geometry' fluctuate and change with time.

"Is there then no *logical* difference between these various systems of statements referred to as geometry?"

None whatsoever. One can prove that if one of these geometries contains a contradiction, then so does each of the others, including Euclidean geometry. In other words, if Euclid's geometry is free from contradiction, then so are the various non-Euclidean geometries. Thus Euclidean geometry is not logically distinguished among the various geometries.

"But can one not decide by observations and measurements which geometry is valid for rods and light rays?"

It certainly will be plausible to you that quantitative differences between the propositions of various geometries concerning drawings, rods, and light rays can be empirically checked only to the degree of accuracy of the measuring instruments. Further reasons why Euclidean geometry is not empirically distinguished, at least in comparison with several non-Euclidean geometries, would take us too far from the subject of our conversation. And even if one geometry could be proved to be valid for objects in the space of our observation the fact would remain that many geometries dealing with undefined lines exist as deductive systems which contradict one another though each of them is consistent; that is, free of contradiction in itself. Again the details would take us too far from our subject. Indeed, I would not have brought geometry into our conversation about ethics at all – as I have told you, the analogy between the two does not seem to me that good – except for the fact that ethical, legal and methodological writings of social scientists include references to geometry – often quite mistaken references, foreign to the spirit of modern geometry. That is why I thought a few remarks might not be useless.

"So there coexist several systems of geometry that contradict each other?"

Yes. And now you see why I say that those who intend to evoke from within themselves a self-evident system of norms for ethical behavior must not be permitted to refer to geometry. Rather, it is the person who establishes the existence of various systems of norms and their possible coexistence, who can speak of analogy to geometry.

(4) "But I fail to see an analogy in the following respect. According to you, the various geometries are incompatible. But as to the various systems of norms for human behavior, I wonder whether it might not be possible, just to a logician, to pool them into one comprehensive system or to distil from them a system contained in all of them; or to construct by some other logical operations a system of norms that disregards special details and thus expresses the essence of morality all the better."

In questions concerning essences I do not consider myself competent, as you already know. A unification by abstraction of various systems of norms into one more general system, however, I regard as a possibility in principle and also as practicable to a certain degree. But consider the following. If some norms of two systems contradict one another, then certainly there cannot exist a consistent third system from which all the norms of both are derivable. For a system from which contradictory rules can be derived is precisely what is called inconsistent or self-contradictory. It is conceivable, however, that one might construct a system that is *compatible* with contradictory norms – of course, at the cost of any content in the subject matter of those norms. A system compatible with contradictory norms concerning certain behavior cannot include *any* regulation of that behavior. Consider, for example, whether in dealing with people, all other conditions being equal, one should or should not take their parentage into account. A moral system simultaneously compatible with norms to favor and not to favor persons of a certain parentage cannot possibly include any concrete precepts regarding this matter. Some such general systems attempt to gloss over this vacuum by decreeing that one should not discriminate between people with *equivalent* parentage. But it is clear that only after being implemented by additional stipulations as to which parentages are to be considered equivalent, and which, if any, superior, does the general system yield any concrete precepts. It thereby becomes contradictory to one of the two norms in question. The extreme case of generality, the claim to grasp the essence of the good, has been discussed in my third note which treats ideals. And the situation is no better when it comes to finding the core shared by various systems of morality. If you try to picture what all imaginable consistent systems of norms for human behavior have in common you will find nothing. There is, in other words, no single norm which would not be lacking in some code or be replaced by a contradictory norm in another. Even if you confine yourself to systems of norms that have been historically or ethnographically observed or preached you will at best find very little that they all have in common. As for other logical operations by which one could develop some kind of *average* of several codes, I can see only one. While there is no average of two contradictory norms such as observance and nonobservance of parentage it may be possible in some cases of codes that differ in several norms to compose an average code by selecting some norms from each, avoiding of course the inclusion of contradictory norms. Such a composition may even be possible in more than one way. But this is probably not what you had in mind as an average of systems of norms.

"No, not exactly. Are the norms of various systems actually so completely contradictory?"

Can you doubt it? It is true, some moral philosophers have sought to obliterate or to conceal or to deny such differences. But for the impartial observer they must be obvious. Is an example necessary besides the one about parentage that I have already mentioned? Monarchical systems grant special privileges to the man born as the first son of the former monarch, regarding that man himself as a monarch. In some systems, certain groups of persons are either favored or disadvantaged on the basis of their parentage: the offspring of aristocrats where there is hereditary nobility; the offspring of the rich where there is inheritance of property; the offspring of Jews or of people with various skin pigmentations where there is racial discrimination. And numerous completely incompatible regulations – conceivable as well as actually existing codes – could be cited regarding sexual and other relationships. Or think of duelling which some codes require, others forbid and still others ignore. The general norms, said to express the will and the normal views of mankind, contain no concrete precepts at all, no matter how great the concern of moral philosophers with those propositions and how much they defend them. Those norms correspond to no actual views or desires of any person or group of persons. Human behavior is regulated precisely the same way whether they are decreed or not. Hence I for my part do not articulate those norms at all.

(5) "But then what motivates a person to accept a definite system of norms?"

An individual's profession of a particular system of norms is an expression of feeling. The actual modeling of his behavior according to some morality is based on a decision.

(6) "As a citizen or as a member of a social class or as a religious believer, an individual is prompted to accept or reject certain norms, above all, by hope of reward and fear of punishment. In matters of tradition, the data of the milieu in which an individual finds himself contribute to his decision. But consider the case described in the second note that you sent me, where the inhabitants of a region, inspired by the best intentions, even renouncing all rights and disregarding petty interests, assemble in order to determine the basic regulation of their existence. There is no question of reward or punishment or, for that matter, of tradition. If under these circumstances an individual suggests or accepts a certain regulation, do you believe that his decision is merely an arbitrary act? Can he not *logically justify* why he opts for a particular regulation of human coexistence when he has a choice between several? Certainly you are aware that in the field of economics, some schools of thought claim to prove logically that the maximum supply and the optimal distribution of commodities and thus the greatest welfare of mankind would take place under certain systems of organization which they describe. What is your opinion of that?"

A logician, after examining the chains of propositions which those economists call logical proofs, might well answer your question with the suggestion that those authors study logic. I, however, am far from overestimating logical proofs in extramathematical matters – although you seem to suspect the opposite – and will not unconditionally say that. I will only suggest that if those economists want to furnish correct proofs, then they indeed must first take the trouble to study logic. This study would without doubt increase their self-criticism. But even aside from the question as to whether or not one can *prove* that particular systems of organization result in certain economic conditions, I ask you which economic conditions are the best. Those where the entire amount of goods present at any time is as large as possible, even if it is distributed in so unequal a manner that envy and conflict ensue? Or where all goods are distributed as uniformly as possible, even if, in comparison to the first distribution, no one wins and many lose? These are just two examples from a large complex of questions each of which can be answered, in the final analysis, merely by an expression of personal taste; for cognitive activity can only state characteristics of the various economic conditions in order to furnish a broad basis for a subjective judgment. But even if there were general agreement with regard to economic goals, and if the means to achieve them were established beyond doubt, what but a subjective judgment can prompt one to choose the system resulting in the *economic* optimum? Whoever accepts as a foregone conclusion that the choice will be made in this direction underestimates the power of ideas – ideas for which, as history demonstrates, not only individuals but often entire groups accept pain and deprivation and suffer the horrors of war: religious ideas, national ideas, tradition, freedom. And to each such idea there corresponds, you must admit, a shaping of the social world, even if you do not admit that an idea *consists* in a proposed shaping of the social world. But choosing between different ideas as well as between different shapes of the world is, in the last analysis, an arbitrary act, a decision. Clearly, the role of cognition is of enormous importance in preparing that act. Let me elaborate on this point. Whoever in pursuing a desire makes a decision without taking into account all forseeable consequences will often find that his decision does not result in the fulfilment of his desire or, if it does, that concomitant conditions materialize which appear to him as very undesirable. He would have preferred the *status quo* to the state of affairs after the realization of his decision, if only he had anticipated the latter state. This particularly applies to decisions about comprehensive, concrete ideas for the shaping of the social world and for the regulation of human relationships. It is true that cognitive efforts in this respect are extremely difficult due to innumerable complications of the subject matter and to important unpredictable elements which usually combine to make results scarce and uncertain. But in view of the paramount importance of such general decisions for the community as well as for the individual it follows that relevant empirical and logical insights should be collected and checked with particular care. In any event, insights of this kind cannot be replaced by vague discourses about ideals and values, though for the sake of such discourses insights are often neglected or ignored. For example, it is a clear and understandable desire of some of the persons with special abilities to let the community benefit from their talents only in exchange for special rewards and, further, to use those rewards in part to secure advantages for their children. Another group of people may have the equally clear and understandable desire to see all children enter the world on an equal footing without any particular advantages or disadvantages. What is gained if either group claims that its desired regulation is the only *just* solution? Yet it may be possible to support either desire by statements about the consequences of its fulfilment for the shape of the social world. Another example: if a group is united by common ancestry and tradition as well as by mental and physical characteristics, then the members may well desire to pass on their tradition and characteristics to later generations. While they would not gain anything by calling their race superior to other races and cross-breeds they might underline their wish by comparing their characteristics with those of other groups, by studying conditions of heredity, and even by a kind of logical argument: to those who say that they presently know too little about characteristics of various races and hereditary conditions to accept the idea, they could respond that keeping their race pure leaves all possibilities open to later generations, no matter what the results of future studies; whereas a cross-breeding now might irretrievably eliminate certain possibilities forever. But, to conclude the discussion of the role of cognitive activities, let me emphasize that a preparation for clear and conscious decisions and for choices between various ideas is all that they can supply. Indeed, even when the question is raised as to whether any regulation of human behavior at all should be imposed – and in view of man's ingenuity and adaptability this question is still more important than the particular form of the regulation – even then cognition can only compare the characteristics of regulated and unregulated situations thereby clarifying what the choice entails. But it cannot prove which of the characteristics are advantageous, and which are disadvantageous. The decision is, ultimately, a matter of taste. It certainly is a fact that in the existing world the vast majority decides in favor of order. Nor can there be any doubt that even if there were no laws providing punishments for violations of order and people were free to choose, perhaps by vote, between an entirely unregulated and a somehow regulated existence the majority would opt for the latter. For the cultural forms that are compatible only with regulated human behavior appeal to most people much more than semi-bestial existence in small unregulated groups concerned only with the protection of life. But the agreement of ever so great a majority, indeed a universal consensus, must not obscure the presence of a subjective judgment and a decision. Precisely when we ignore the fact that existing laws and penalties make the choice for each individual easy; precisely in the case you mentioned where all convened of their own accord to regulate their existence without fear of punishment or hope of reward, it should become clear that it is not insight but rather a decision of the individual that is the ultimate reason for his attitude – not cognition but volition.

(7) "By no means, however, can you ignore the role of *faith* which is seen to be of such great importance in the adoption of moral systems. I would have expected you to mention religious ideas as a basis of morality when discussing ideals; and to add critically that the statement "Good is what corresponds to the will of God" only receives content when it is explained in what the will of God consists, noting that even related monotheistic religions do not agree in their explanations. Since you did not treat religious ideas in your third note I conclude that you recognize an epistemological difference between them and all the other ideals that you criticized. In basing precepts on the will of nature or on justice one must *decree* in what these ideals consist, whereas each religion includes by *tradition* a system of commandments and prohibitions handed down as the will of a superhuman being to whom the world and mankind owe their existence. The believer in a religion need not establish facts about the will of God; he considers these as given by revelation and obeys the prescribed commandments as an expression of his *faith* – incidentally not only from fear of punishment but also out of love for the law-giver and for the forefathers of his faith."

Actually, the believer has a very complete, self-contained world view. Just from a purely logical point of view, this fact is usually too little emphasized. A few basic assumptions, especially the revelations that you have mentioned, supply him with a world view that includes moral precepts. The factor of will that I have characterized as the basis for the acceptance of a moral system, however, is not lacking even in the case of a believer. Logical analysis demonstrates that it is transferred, consciously or unconsciously, into the acceptance of the basic assumptions; that is, into the act or, perhaps better, the state of faith.

(8) "This interpretation will probably also be acceptable to the believer. But from the point of view of scientific ethics, I am not satisfied with your remarks. Your principles seem to result in a demystified but *superficial* ethics. All you want to examine is what determines actual events, and this is why you limit yourself to the study of decisions. For example, in your criticism of the categorical imperative, you deal only with its concrete consequences; but with Kant the concept of duty includes not only acting *according to the law* but, above all, acting *for the sake of the law*; and this you ignore. You ignore it consciously, since you state explicitly that you do not care to discuss the questions that interested Kant himself and other philosophers. In this way, however, your ethics relegates *intent* to the background and disregards it for the sake of actual *decisions*. Thus my main objection to your ethics is that you do not seek to understand *morality* but rather to examine the *will* in order to understand *the shaping of the social world* or possibly *the formation of human associations*. But is the basis of your entire limitation *itself* anything but a judgment of taste? Is not its only source an arbitrary decision concerning scientific research? I do not wish to elaborate on this remark, however, for I have another objection based on your own principles, which may therefore affect you more directly. Throughout the entire exposition of your ideas, with its contempt for disputes over terms, for all definitions, and for delimitations of ethical concepts, there seems to run a certain *logical pragmatism*. I am referring to your opinion that the only justification for scientific definitions – or, as you may say, for the way words are used in science – lies in their *fertility*; that is, in the possibility of using them in a great number of important propositions, preferably practical propositions."

This indeed is very definitely my opinion. Whoever delimits concepts without forseeable consequences is like a man dividing up the desert into lots.

"Fine. But what can the application of exact thought to questions of ethics achieve? Obviously you must go beyond describing the origin of the words 'good' and 'evil' and establishing the fact that the actual acceptance of a particular morality is based on a decision?"

I certainly cannot claim to be in a position to make *many* ethical statements – in the same way, incidentally, as I am not aware of *many* statements in traditional ethics after the mode of expressing them has been logically purified and made concise. But certainly I believe that I am able to go beyond what you have just mentioned.

"In what sense or direction? So far, I know only your views concerning the relationship between ethics and geometry."

(9) I believe that I am able to move toward a *logic of ethics.*

"A logic of ethics? Do you want to remind us that the duty to behave in a certain way is the same as the prohibition of the opposite behavior and implies that the commanded behavior is permitted? That a forbidden behavior, therefore, is not one's duty? But all this is perfectly obvious."

Given a behavior *B*, consider the four propositions:

1. *B* is commanded, 2. *B* is not commanded,
3. *B* is permitted, 4. *B* is not permitted.

Of their six conjunctions in pairs, three are absurd, namely '1 and 2,' '3 and 4,' '1 and 4'; two are equivalent with single propositions, namely '1 and 3' with '1,' and '2 and 4' with '4.' But '2 and 3' leads to what may be called *optional* or *discretionary* behavior – permitted but not commanded. Clearly, the entire situation is analogous to that prevailing in the logic of modalities which distinguishes necessary and nonnecessary, possible and nonpossible propositions. Discretionary behavior is analogous to accidental (possible but not necessary) propositions. Once correctly formulated, all such elementary formal statements do indeed seem obvious. But apart from the logic of norms, I have in mind quite different formal statements which have not to my knowledge ever been correctly formulated.

(10) "I imagine you are thinking of a logic of desires – a theory consisting of propositions such as 'If one desires *A* and *B*, then he desires *A* and desires *B*,' and vice versa."

Careful! Note two points. *First*, propositions about desires can be connected in the same way as any other propositions, in particular, by means of the particles 'and,' 'or,' 'if... then,' 'not.' Compounds with the constituents 'I wish a pencil' and 'I wish a pen' include, for example, the propositions 'I wish a pencil and I wish a pen,' 'I wish a pencil or I wish a pen,' 'I do not wish a pencil.' Moreover, the truth or falsehood of such a compound proposition depends upon the truth and falsehood of its constituents in the same way as does any other complex proposition. 'I wish a pencil and I wish a pen' is true if and only if both 'I wish a pencil' and 'I wish a pen' are true. Of the two propositions 'I wish a pencil' and 'I do not wish a pencil' exactly one is true – not more than one, according to the law of contradiction; at least one, according to the law of the excluded middle. *Second*, the possible objects of a person's desires are connected in the same way as any other objects. A pen and a pencil, as well as a pen or a pencil is a possible object of desire. So is the nonpossession of a pen or the nonoccurrence of an event.

"If the logic of desires is to teach us anything beyond the generally valid laws of logic, then it probably can do this only by establishing regular relationships between the connections of propositions about desires and the connections of the objects of desires; for example, the equivalence that I mentioned of the propositions 'I desire *A* and I desire *B*' and 'I desire *A* and *B*.' But are not such statements logical truisms?"

By no means. That is why I cautioned you. Whether or not this equivalence is valid is not a logical question; only experience can decide. And experience shows that the equivalence is *not* in general valid. In the first place, it is entirely possible that the proposition 'I desire *A* and I desire *B*' is true while the proposition 'I desire *A* and *B*' is false. I may wish to be invited by my friends *X* tonight and I may wish to be invited by my friends *Y* tonight. But to be invited tonight by the *X*'s and the *Y*'s may not be desirable to me; in fact, it may merely place me in an embarrassing situation. Secondly, it is entirely possible that 'I desire *A* and *B*' is true and yet 'I desire *A* and I desire *B*' is false or even that 'I do not desire *A* and neither do I desire *B*' is true. If, on a hike, I wish to record a thought, then I desire paper and a writing utensil. But I do not desire paper by itself, that is, without a writing utensil, nor a writing utensil by itself. These and similar facts are special cases of a phenomenon known in economics as complementarity of goods: certain objects are desired in combination but not individually. Incidentally, the lack of regularity governing connections between compound propositions about, and complex objects of, desires is not confined to conjunctions; the same problem exists with regard to disjunctions and negations. There is no general rule relating the propositions 'I desire *A*' and 'I desire non-*A*.' Regardless of whether the first proposition is true or false, the second may be either true or false. While according to the law of contradiction it is impossible for a person to desire *A* and not to desire *A* it is entirely possible that a person desires *A* and desires non-*A*, if perhaps with different intensities. For example, some evening I may wish to stay home in order to work, and I may wish not to stay home in order to attend a concert. Similarly, whereas according to the law of the excluded middle either I do desire *A* or I don't desire *A*, it can happen that I neither desire *A* nor desire non-*A*, though my aversions may be of different intensity. You see that general relationships between compound propositions about desires and complex objects of desires do not exist – certainly not in the simple form that you seem to envisage. This lacuna is not due to a shortcoming of logic. It is caused by the complication of the empirical material, that is, of the relationships that you have mentioned. By the way, I wish that the same degree of precision in using the logical particles that we have found indispensable in dealing with desires would be exercised throughout the humanities. But the use of those logical connectives in the social sciences only too frequently offers an unpleasant spectacle to the logician; 'therefore' and 'consequently,' in particular, are often used by social scientists as though they weie emotion-charged interjections.

"You still owe me an indication of what a logic of ethics can achieve. Do any interesting problems arise from the fact that the objects of desires may themselves be desires? You hinted at that possibility in your first note when mentioning the voice of conscience. But take such simple propositions as 'I wish that all my wishes come true' or 'I wish I had a wish' or 'I wish that none of my wishes come true.' Obviously, in each of these sentences the word 'wish' is used, as it were, on two different levels. To ignore this fact would probably lead one to contradictions."

That is correct. For example, suppose someone were to ask me the following three questions: "Do you wish to lose your pen? Do you wish to break your leg? Do you wish that none of the wishes about which I am asking be fulfilled?" Naturally, I can immediately answer the first two questions negatively. But the third I can answer neither affirmatively nor negatively. For assume first that I answer "No" or, in other words, that

I say "I do not wish that none of the three wishes be fulfilled." This would be tantamount to saying "I wish that at least one of the three wishes be fulfilled." Since it is neither of the first two it must be the third that I wish to be fulfilled. Thus my answer "No" implies the answer "Yes" to the third question – a contradiction, which rules out the answer "No." But assume that I answer "Yes" to the third question. This would in particular imply that I do not wish the third wish to come true; in other words, it would imply the answer "No" – again a contradiction. That I neither do nor do not wish the fulfilment of the third wish contradicts the law of the excluded middle. But it is a paradox following step by step a well-known logical paradox and, therefore, can be clarified in an analogous way. Given a class of wishes, which may be called wishes of type 1, suppose one also discusses wishes about these wishes, which may be called wishes of type 2, and wishes about wishes of type 2, which may be called wishes of type 3, and so on. Then, whenever one speaks of *all* wishes, one must add *of which type*. If in the example of the three wishes that I have mentioned the first two are of type 1, then the third is of type 2; and the third question may be correctly reformulated as follows: "Do you wish that none of the wishes of type 1 about which I am asking be fulfilled?" This question I can of course answer affirmatively while answering the first two questions negatively. The paradox is the result of confusing types, since the word 'all' in the original third question refers to the two wishes of type 1 and the wish of type 2 mentioned in that third question. Thus a theory of types of desires, which avoids paradoxes that otherwise appear, would constitute one chapter of the logic of desires. But it would follow Russell's theory of types in logic; and so would hardly offer anything substantially new.

"This is just what I object to: the poverty of the demystified ethics. Logical thought, wherever used cathartically, can be expected to provide a good deal of critical purification. But, above all, I expect it to provide positive results, especially when used by one whose mathematical achievements are reputed to be so constructive. And since in your five notes you have criticized the writings of well-known philosophers because they lack positive, and especially practical, conclusions you cannot dodge the demand for positive results of scientific thought in your demystified ethics. I admit that your critique reveals the important role of decisions; but obviously decisions of various individuals may conflict with each other. Will exact thinking make a positive contribution to the basic problem how under these circumstances human behavior ought to be regulated? I am against the application of logic to questions of morality for the same reasons that make you reject delimitations of certain ethical concepts, of the word 'science,' or of the objects of particular sciences: I feel that the development of an ethics according to your principles may be so sterile that it is not worth the effort."

My dear friend, with your last question you have become untrue to yourself. Don't you remember warning me that I should not be misguided into formulating subjective evaluations in ethics which strives for objective knowledge? Accordingly, I have directed all my efforts, in the notes that I sent you, as well as in this conversation, to the establishment and ordering of facts. Many of these facts concern the presence and the effects of feelings and arbitrary acts in matters of morality. And even if, precisely for that reason, I found it difficult to keep from expressing my own opinions, I have refrained; and you certainly were never disturbed by the intrusion of my personal feelings or evaluations. But now you ask how human behavior ought to be regulated. According to my principles, as a philosopher of ethics I must limit myself to the study of how regulations are *actually* established, and to the examination of how they *can be* established, concentrating my efforts naturally on possibilities that may have escaped you or of which you may not be completely aware. You know that I interpret the more general question of what ought to happen either as a question about what is in accordance with certain norms or as a question about the wishes of the person asked or the wishes of some third party. Your question cannot have the first meaning since that which ought to happen is precisely the creation of norms. My personal desires, feelings, or evaluations, however, I cannot inject into our ethical discussions in observance of your previous warning. For now, therefore, I must leave your question unanswered. But I'll send you a second series of notes, which will show you more clearly what I mean by applying exact thinking to questions of ethics. It is not what you seem to think it is – not a logic of norms or desires. The five notes that I will send you rather deal with human groups and associations – and by no means in a purely critical or negativistic spirit. The notes discuss the relations between groups and between groups and individuals; and these relations lend themselves to positive statements.

CHAPTER IV

FIVE LOGICO-MATHEMATICAL NOTES ON VOLUNTARY ASSOCIATIONS

1. THE PARTITIONS OF PEOPLE INDUCED BY NORMS

Many norms originated in prehistoric times. Some of them demand actions or omissions that may originally have been instinctual; others, their origin hidden in obscurity, are presumably the result of mere coincidences. Later, further norms were created, while older rules were changed or abolished by powerful individuals and groups who were interested in augmenting their own or their descendants' power.

Norms were combined into systems. For the most part, the norms in historical systems are logically independent of one another; that is, not only each norm but also the negation of each norm is compatible with the conjunction of the others. In a strict logical sense the norm "Thou shalt not steal" is compatible with "Thou shalt not covet thy neighbor's goods" as well as with "Thou shalt covet thy neighbor's goods."

The historical ties, however, between some logically independent norms are very strong; for example, between certain regulations of sexual behavior and certain regulations of the inheritance of property. In many cases, historical connections are purely accidental; in others they stem from the (erroneous) belief that various norms are logically derived from a more general rule or from an ideal. Due to the lack of an actual content of the more general rule, a logically *correct* derivation of the particular norms is out of the question, the general rule or the ideal being also compatible with the opposite of the particular norm. This is quite apparent in norms that are "deduced" from the postulate of justice or the demand for a life in accordance with nature.

Each particular norm or system of norms, be it prehistoric in origin or decreed by power or introduced by reformers or due to arbitrary conventions, has a group of adherents; that is, of people who follow the norm or the system. This group changes with time. It is very large if the norm is a law the violation of which is subject to punishment or if the norm is actually desired by a large number of people. The group may shrink if compliance with the norm produces consequences contrary to the desires of many persons – consequences that may not have been present or were not anticipated when the norm was formulated. The group may also change because human relationships appear or disappear with changing times and conditions, above all, with technological progress. If an inventor were to succeed in bringing about a situation in which everyone could avail himself without effort of all goods in any desired quantity, then the norm "Thou shalt not steal" would be pointless.

Thus every single norm and system of norms has a history mirrored in the size of the group of its adherents and changes in it. But this history of a norm is less typical than the development of an individual's moral judgments; that is, the histories of various norms display fewer common traits and regularities than are found in the unfolding of a man's judgment of various types of behavior. These judgments, reflected in his classification of actions as good, indifferent, and evil, are fairly regularly connected with his education and with general factors of life.

At any rate, in analogy to the tripartition of the modes of human behavior in the moral judgments of each rational individual, there is a tripartition of people in connection with each comprehensible norm. Each class of individuals is divided into the group of those who approve of the norm, the group of those who are indifferent to it, and the group of those who disapprove of it. Of course, in a given class of human beings, one of these groups may be empty; there may be even only one nonempty group, which is the case if the attitudes of all members of the class coincide.

Because of a gradation of interest in various norms, there are in many cases only blurred dividing lines between the group of indifferent individuals, on the one hand, and the groups of approving or disapproving persons, on the other. Moreover, corresponding to the three types of evaluations (according to actions, words, and desires, cf. p. 8) a norm in fact gives rise to three tripartitions of each class of individuals. People are divided into those who always follow, or sometimes do and sometimes do not follow, or never follow the norm in their actions; those who agree with or are indifferent to or disagree with the norm in their verbal utterances; and those who agree with or are indifferent to or disagree with the norm in their inner desires. But in what follows, for the sake of simplicity, we will disregard blurred dividing lines as well as possible differences between these three tripartitions confining ourselves instead to the study of the case where they are identical.

The two tripartitions corresponding to two different norms are in many cases disparate. Among the people approving of the first norm there may be some approving of, some indifferent to, and some disapproving of, the second; similarly, among the people indifferent to, and the people disapproving of the first norm. A system consisting of two norms, therefore, gives rise to nine possible attitudes and hence to a division of each class of people into nine or fewer nonempty groups each consisting of all those whose attitudes toward both norms coincide. There are fewer than nine nonempty groups if some of the possible postures are not taken by anyone in the class. If the first norm forbids murder for the sake of robbery, then whatever the second norm, the nine groups reduce to at most three. Since everyone is in favor of the first norm, at least six of the nine groups are empty.

In the same way, a system of three norms gives rise to twenty-seven attitudes and to a division of any class of people into twenty-seven or fewer nonempty groups. More generally, if a system consists of n norms, then there are 3^n possible attitudes – a very large number if n is large – and each class of people is divided into 3^n or fewer nonempty groups, each consisting of all those whose attitudes coincide. We shall call them *total groups of consentience with regard to the n norms*. These groups are (1) jointly exhaustive: each member of the class belongs to one of them, and (2) mutually disjoint: no two of them have a member in common. Hence each member of the class belongs to one and only one of the total groups of consentience. A division of a class into jointly exhaustive and mutually disjoint groups will be referred to as a *partition* of the class.

The number of human relationships to be regulated is large. The number of total groups of consentience with regard to *all* regulating norms – briefly, of the *total groups of consentience* – may be very large, while the size of many of those groups may be small. Indeed, many a person, P, finds only a few individuals, if any, who *completely agree* with him; that is, whose attitudes about all norms coincide with his own. If there is no such individual, then P constitutes a total group of consentience by himself. If more than one individual completely agrees with P, then any two of these individuals also completely agree with one another. This property of the agreement relation is referred to as *transitivity*: two individuals who completely agree with a third also completely agree with one another. Hence the individuals who completely agree with P consitute one of the jointly exhaustive and mutually disjoint total groups of consentience; this group is said to be *determined* by P and is called *P's total group of consentience*. If P_1 is any member of this group, then the group determined by P_1 is the same as the group determined by P.

A simple procedure for partitioning a class into total groups of consentience is the following: select any member, P_1, of the class and form the group determined by P_1. Unless all members of the class completely agree, select an individual P_2 who does not, and form P_2's total group of consentience. Unless all members of the class completely agree with either P_1 or P_2, select an individual, P_3, who does not, and form P_3's total group of consentience. Proceed in this way until no one is left. By then, each member of the class will have been assigned to one of the groups; that is, the partition of the class into total groups of consentience will be complete. This partition procedure is independent of the particular individuals selected, except for the order in which the groups are produced; but the order is irrelevant. For example, if P_2 were selected first and P_1 second, then the two initial groups would be obtained in reverse order, but have the same membership as before. Or if instead of P_1 another individual from P_1's total group of consentience were selected, he would determine, as has already been observed, the same total group of consentience as did P_1.

A class of individuals may also be partitioned into groups according to the condition that *only* persons who completely agree belong to one group without it being required that *all* completely consentient individuals be included in the group. We then speak of *groups of consentience* – total or nontotal. Obviously, each group of consentience is contained in precisely one total group of consentience while the latter may contain several (nontotal) groups of consentience. Clearly, of all partitions of a class into groups of consentience, the one composed of the least number of groups is the partition into *total* groups.

A significant reduction in the number of groups and an increase in their size can in general be achieved by partitioning a class into *groups without dissension*. A group is said to be without dissension if no two of its members dissent from one another about any norm in the sense that one approves and the other disapproves of it. A person who is indifferent to a norm may be in the same group without dissension as an individual who is either approving or disapproving of the norm provided the two agree in their attitude toward all norms to which neither is indifferent. Each (total or nontotal) group of consentience is of course also a group without dissension.

In contrast to the agreement relation, the relation of not dissenting fails to be transitive. To illustrate this fact by an extreme example, let P_0 be some person who is indifferent to *all* norms and consequently dissents from no one. While P_0 dissents neither from an individual approving of a particular norm nor from an individual disapproving of that norm the latter two individuals dissent from one another. (What can be said about all individuals not dissenting from a person, P, is merely this: they do not dissent from each other about those norms of which P approves or disapproves.) It follows that the group of all individuals not dissenting from one and the same individual is not in general a group without dissension.

While uniting with an individual, P, all people not dissenting from P does not necessarily form a group without dissension the following procedure does. Select an individual, P', not dissenting from P; then an individual, P'', not dissenting from either P or P'; then an individual not dissenting from any one of the three individuals already selected, and so on, as long as desired or as long as possible. This procedure must stop when everyone who is left dissents from at least one of the group about at least one norm, even though there may remain some individuals not dissenting from P. At this point, one has formed a *maximal* group without dissension; that is, one that cannot further be extended. And it is clear that each individual is a member of at least one maximal group without dissension. (Each individual by himself constitutes a *minimal* group without dissension.)

It should be noted that the maximal group without dissension including P obtained by the formation procedure described may well depend upon which individuals P', P'',... have been selected in the course of the procedure. For there may be several maximal groups without dissension including P. This is particularly obvious in the case of a person P_0 who is indifferent to all norms. Let P_1 and P_2 be two individuals one of whom approves of some particular norm while the second disapproves of it. The procedure of forming a maximal group without dissension including

P_0 leads to different groups depending upon whether P_1 or P_2 is selected as P'.

Each class of people can be partitioned into (disjoint) groups without dissension – most classes in several ways. For example, select any member, P_1, and form a maximal group without dissension including P_1. Unless the entire class is without dissension there are people not belonging to that group. From among them, select a person, P_2; and applying the formation procedure described above to the class of people outside of the first group extract from them a second group without dissension – a group including P_2. Continue the procedure to the point where only those remain who dissent from at least one member of the second group and, of course, also from one of the first. This second group without dissension is maximal relative to the people outside of the first group. (But note that it is not necessarily maximal in the total class, since the first group may include members not dissenting from any member of the second.) If there are people outside of both groups formed so far, then select one of the outsiders, P_3; and applying the formation procedure to the class of all the outsiders form a group without dissension including P_3 and maximal relative to the class of the outsiders. Proceed in this way until no one is left, at which point the entire class is partitioned into groups without dissension.

If no one is indifferent to any norm, then the partition obtained is the (unique) partition into total groups of consentience and all of them are maximal groups without dissension. In general however, it is clear (1) that the groups without dissension obtained by the foregoing procedure, except the first, need not be maximal, and (2) that their composition may depend upon which individuals $P_1, P_2, P_3, \ldots$ have been selected in the course of the partition procedure. This again is especially obvious if some individual P_0 is indifferent to all norms. For P_0, dissenting from no one, belongs to each maximal group without dissension, in particular to the group including P_1; but since the groups of the partition are mutually disjoint, P_0 is not a member of the other groups which consequently are not maximal. Hence a partitioning into maximal groups without dissension is impossible in this case. Had P_2 rather than P_1 been chosen first, then P_0 would be a member of the maximal group including P_2, and at least two groups in the second partition would be different from all groups in the first.

We thus see that whereas each class can be partitioned in one and only one way into total groups of consentience, some classes admit more than one partition into groups without dissension and some admit no partition at all into maximal groups without dissension.

Let G be any of the groups without dissension resulting from the partition described above, and consider a member P of G. If P' completely agrees with P, that is, if P' belongs to the total group of consentience determined by P, then, as one readily sees, P' is also a member of G. Hence G includes the total group of consentience of each member of G. Consequently, each total group without dissension resulting from our partition is either a total group of consentience or the union of such groups. It follows that the number of those groups without dissension is less or at any rate not greater than the number of total groups of consentience; while the membership of the former groups is greater or at any rate not less than that of the latter.

This seems to be a particular advantage of the described partition of a class into groups without dissension. But one may prefer other points of view. For example, it may be desirable to form groups without dissension of roughly equal size. In this case, members who are indifferent to a norm will be used not to maximize the initial groups, but to equalize the size of all of them.

In view of the gradation of people's feelings toward norms one may achieve a still further reduction of the number of groups and a further increase in the membership of the groups by forming *groups without important dissension.* By such a group, we mean a group without dissension in which those for whom the norm is of minor importance declare themselves indifferent to it. Such a group may even include a person approving and a person disapproving of the same norm provided that at least one of them does not attach great importance to its substance. Partitions thus obtained depend upon the degree of tolerance existing in the class. The extreme case of no tolerance on the part of those who are not indifferent leads to the groups without dissension. In the other extreme case of complete tolerance, the entire class is one group without important dissension. Of course, also the number of *partitions* into groups without important dissension in general increases with increasing tolerance. This makes it easier to equalize the size of the groups, where this appears desirable.

2. Duality

The preceding note has dealt with the class of people approving of a norm. The object of the first of the epistemological notes was the class of norms approved of by an individual. This parallelism is not accidental. The deeper reason for it is a duality which occasionally is of heuristic significance and will now be examined more closely. This study is preceded by a few remarks about some simple set-theoretical concepts used throughout the remainder of this chapter.

We denote sets by capital Gothic letters. If $\mathfrak{A}$ and $\mathfrak{B}$ are two sets, then we are often interested in

the *union* of $\mathfrak{A}$ and $\mathfrak{B}$, denoted by $\mathfrak{A} \cup \mathfrak{B}$,

which is the set consisting of all elements belonging to at least one of the two sets; we also say that $\mathfrak{A} \cup \mathfrak{B}$ is the class of the individuals belonging to $\mathfrak{A}$ *or* $\mathfrak{B}$ ('or' being used in the sense of the legal 'and/or' and the Latin '*vel*');

the *intersection* of $\mathfrak{A}$ and $\mathfrak{B}$, denoted by $\mathfrak{A} \cap \mathfrak{B}$,

which is the set of all elements belonging to both sets; we also say that $\mathfrak{A} \cap \mathfrak{B}$ is the set of the individuals belonging to $\mathfrak{A}$ *and* $\mathfrak{B}$.

These concepts can be extended to more than two sets. For example, if $\mathfrak{A}_1, \mathfrak{A}_2, \mathfrak{A}_3$ are the populations of Canada, the United States, and Mexico, then $\mathfrak{A}_1 \cup \mathfrak{A}_2 \cup \mathfrak{A}_3$ is the population of North America. If $\mathfrak{B}_1$ is the population of Asia and $\mathfrak{B}_2$ that of the Soviet Union, then $\mathfrak{B}_1 \cap \mathfrak{B}_2$ is the population of Siberia. If at some instant, $\mathfrak{C}_1$ is the population of America and $\mathfrak{C}_2$ that of Europe, then $\mathfrak{C}_1 \cap \mathfrak{C}_2$ includes no member; it is the *empty* set, denoted by $\varnothing$; in symbols, $\mathfrak{C}_1 \cap \mathfrak{C}_2 = \varnothing$. If the intersection of two sets is empty the sets are said to be *disjoint*.

The only sets studied in the sequel consist either of men or of norms. The totalities of men and of norms considered are referred to as the *classes* $\mathfrak{M}$ and $\mathfrak{N}$. Subsets of $\mathfrak{M}$ and $\mathfrak{N}$, sets of men or norms with certain properties, will be referred to as *groups* of men or norms.

If M is an individual belonging to $\mathfrak{M}$, and N a norm belonging to $\mathfrak{N}$, then we assume that one of the following three situations obtains:

M *approves* of N, M is *indifferent* to N, M *disapproves* of N.

We shall characterize the three possible attitudes by the symbols $+$, 0, $-$, and also write

$(M,N)=1, \quad (M,N)=0, \quad (M,N)=-1,$ respectively.

For the sake of simplicity, we thus disregard several possibilities: that M's judgments about the norm N deviate from one aonther according to his thoughts, words, and deeds; also that M takes no position at all with regard to N instead of assuming one of the three indicated attitudes (postulating, as it were, a law of the excluded fourth).

If M is any individual belonging to the class $\mathfrak{M}$, then the class $\mathfrak{N}$ of all norms considered is partitioned into three jointly exhaustive and mutually exclusive groups

$\mathfrak{N}_M^+$ of all norms approved of by M;
$\mathfrak{N}_M^0$ of all norms to which M is indifferent;
$\mathfrak{N}_M^-$ of all norms disapproved of by M.

Clearly,

$$\mathfrak{N} = \mathfrak{N}_M^+ \cup \mathfrak{N}_M^0 \cup \mathfrak{N}_M^- \quad \text{and}$$
$$\mathfrak{N}_M^+ \cap \mathfrak{N}_M^0 = \varnothing\,, \quad \mathfrak{N}_M^+ \cap \mathfrak{N}_M^- = \varnothing, \quad \mathfrak{N}_M^0 \cap \mathfrak{N}_M^- = \varnothing\,.$$

Since the union of empty sets is empty the last three formulas can be combined into one, namely

$$(\mathfrak{N}_M^+ \cap \mathfrak{N}_M^0) \cup (\mathfrak{N}_M^+ \cap \mathfrak{N}_M^-) \cup (\mathfrak{N}_M^0 \cap \mathfrak{N}_M^-) = \varnothing\,.$$

The dual counterpart of the tripartition of the class $\mathfrak{N}$ determined by an individual M is a tripartition of the class $\mathfrak{M}$ of all people under consideration determined by a norm N into the jointly exhaustive and mutually exclusive groups

$\mathfrak{M}_N^+$ of all people approving N;
$\mathfrak{M}_N^0$ of all people indifferent to N;
$\mathfrak{M}_N^-$ of all people disapproving of N.

From the assumptions it follows that

$$\mathfrak{M} = \mathfrak{M}_N^+ \cup \mathfrak{M}_N^0 \cup \mathfrak{M}_N^- \quad \text{and}$$
$$(\mathfrak{M}_N^+ \cap \mathfrak{M}_N^0) \cup (\mathfrak{M}_N^+ \cap \mathfrak{M}_N^-) \cup (\mathfrak{M}_N^0 \cap \mathfrak{M}_N^-) = \varnothing\,.$$

An individual, M, belongs to the group $\mathfrak{M}_N^+$ if and only if the norm N belongs to the group $\mathfrak{N}_M^+$. Similarly, N belongs to $\mathfrak{N}_M^0$ or $\mathfrak{N}_M^-$ depending upon whether M belongs to $\mathfrak{M}_N^0$ or $\mathfrak{M}_N^-$. We join these three statements to form the following:

$$M \text{ belongs to } \mathfrak{M}_N^* \text{ if and only if } N \text{ belongs to } \mathfrak{N}_M^*,$$

where $*$ stands for $+$ or 0 or $-$.

Now let $\mathfrak{M}$ be a class of men, $\mathfrak{N}$ a class of norms, M_1, M_2 two members of $\mathfrak{M}$ and N_1, N_2 two members of $\mathfrak{N}$. Since, by definition, (M,N) is either 1 or 0 or -1, the formula $(M_1,N)=(M_2,N)$ indicates that (M_1,N) and (M_2,N) are either both 1 or both 0 or both -1. The product of two of the numbers 1, 0, -1 is 0 or 1 unless one factor is 1 and the other is -1, in which case the product equals -1. Hence $(M_1,N)\cdot(M_2,N)$, the product of the numbers (M_1,N) and (M_2,N), equals -1 if and only if

$$\text{either } (M_1,N) = 1 \text{ and } (M_2,N) = -1$$
$$\text{or } (M_1,N) = -1 \text{ and } (M_2,N) = 1\,.$$

A combination of these statements with their duals can be tabulated as follows:

	M_1 and M_2 judge N	N_1 and N_2 are judged by M
In the *same* way	$(M_1,N)=(M_2,N)$	$(M,N_1)=(M,N_2)$
In *opposite* ways	$(M_1,N)\cdot(M_2,N)=-1$	$(M,N_1)\cdot(M,N_2)=-1$

Using these formulas one can define consentience, dissentience and related concepts as well as their duals.

If $(M_1,N)=(M_2,N)$ holds

(a) for each N in $\mathfrak{N}$, then the individuals M_1 and M_2 are *consentient* with respect to the class $\mathfrak{N}$;

(b) for each pair M_1, M_2 of individuals in $\mathfrak{M}$, then the norm N is said to *unify* the class $\mathfrak{M}$;

(c) for each N and each pair M_1, M_2, then the class $\mathfrak{N}$ is said to *unify* $\mathfrak{M}$.

Dually,

if $(M, N_1)=(M,N_2)$ holds

(a′) for each M in $\mathfrak{M}$, then we say that the norms N_1 and N_2 are *conjunct* with respect to $\mathfrak{M}$;

(b′) for each pair N_1, N_2 of norms in $\mathfrak{N}$, then we say that M *codifies* the class $\mathfrak{N}$;

(c′) for each M and each pair N_1, N_2, then the class $\mathfrak{M}$ is said to *codify* $\mathfrak{N}$.

In case (b), either all members of $\mathfrak{M}$ approve of N or all are indifferent to N or all disapprove of N. In case (c), the various norms are not necessarily judged in the same way: some may be approved of by all members of $\mathfrak{M}$, others disapproved of by all of them, and still others meet with general indifference. Dually, in case (b′), M either approves of or is indifferent to or disapproves of all norms in $\mathfrak{N}$. In case (c′), the various members of $\mathfrak{M}$ do not necessarily judge in the same way: some may approve of all norms in $\mathfrak{N}$, others disapprove of and still others be indifferent to all of them.

Similarly,

if $(M_1,N)\cdot(M_2,N)\neq -1$ holds

(d) for each norm N in $\mathfrak{N}$, then M_1 and M_2 are *non-dissentient* with respect to $\mathfrak{N}$;

(e) for each pair M_1, M_2 of members of $\mathfrak{M}$, then N is said to be *non-polarizing* with respect to $\mathfrak{M}$;

(f) for each N and each pair M_1, M_2, then $\mathfrak{N}$ is said to be *non-polarizing* with respect to $\mathfrak{M}$.

Dually,

if $(M,N_1)\cdot(M,N_2)\neq -1$ holds

(d′) for each member M of $\mathfrak{M}$, then the norms N_1 and N_2 are said to be *non-disjunct* with respect to $\mathfrak{M}$;

(e′) for each pair of norms N_1, N_2 in $\mathfrak{N}$, then M is said to be *undivided* on $\mathfrak{N}$;

(f′) for each M and each pair N_1, N_2, then $\mathfrak{M}$ is said to be *undivided* on $\mathfrak{N}$.

A group unified by $\mathfrak{N}$ is a group of consentience with respect to $\mathfrak{N}$; a group not polarized by $\mathfrak{N}$ is a group without dissension.

The duality studied in this note cannot be applied to relations between persons and norms beyond those symbolized by $+$, 0, $-$. For example, two logically equivalent norms (either one implying the other) are conjunct; but only the conjunction of two norms has a dual counterpart, namely, the consentience of two individuals. Logical connections of norms such as equivalence or implication or contradiction, cannot be dualized in a strict sense. Yet in some instances the idea of duality affords at least a suggestion as to where counterparts of some kind may be sought. What are roughly analogous to logical relations between norms are relations between individuals based on power, coercion, common descent or the like, which unify or separate individuals in their attitudes toward norms. But such relations between individuals include empirical elements, which are absent from logical relations between norms.

Also the following is perhaps worth mentioning. Even in situations which can be expressed in terms of the attitudes $+$, 0, $-$ and which, therefore, lend themselves to dualization in the strict sense, there may exist, in addition, the possibility of more material dualizations. For example, the idea of an individual ranking norms according to the various degrees of interest he takes in them can be strictly dualized. If *M* is an individual dividing norms simply into those he considers important and those he considers unimportant, then the dual counterpart, if *N* is a norm, is the division of people into those who consider *N* important and those considering *N* unimportant. Dual in an, as it were, more material sense is the ranking of individuals by a norm according to their importance as actually effected by some norms. In contrast to strict duality such analogies are rather vague. Dualization thus not only leads to clarification but its applicability is often an indication of clarity already achieved.

3. Disjunctive norms

In the preceding note, inferences have been drawn from the assumption that every individual has one of three attitudes toward a norm: approval, indifference, or disapproval. But disapproval of a norm is something rather indefinite since it is compatible with the approval of a variety of other norms. The present note is based on more detailed assumptions.

In what follows, the frame of reference consists of three elements: a class, $\mathfrak{M}$, of individuals; a certain conduct or situation, α; and a class, $\mathfrak{N}$, of norms proposed for the regulation of α among the members of the class $\mathfrak{M}$. A class $\mathfrak{N}$ consisting of k norms, $N_1, N_2, \ldots, N_k$ will be referred to as a class $\mathfrak{N}_k$. It will be assumed that the norms belonging to $\mathfrak{N}$ exhaust the possible regulations of α. Consider, for example, the case where $\mathfrak{M}$ is the population of a city, α is the conduct on Sunday mornings, and the class of proposed norms is $\mathfrak{N}_3$ consisting of the following precepts: to attend religious services, to further one's scientific or artistic education, to practice physical exercise. Within this frame of reference no other regulation will be considered. If another norm were to be taken into account (say, complete relaxation), then the frame of the theory could be, and would have to be, extended. If it is also proposed to leave the conduct or situation α altogether unregulated, then for some purposes it is convenient to introduce the content of this proposal as a special norm, N_0. But at the end of this note it will be demonstrated that the role of N_0 is different from that of the other norms.

We first assume that each member of $\mathfrak{M}$ is willing to accept some of the norms belonging to the class $\mathfrak{N}$ as precepts for everyone and is not willing to accept the other norms. As an example, let $\mathfrak{N}$ be a class $\mathfrak{N}_3$ consisting of three norms, N_1, N_2, N_3. Then the class $\mathfrak{M}$ can be divided into seven groups: the group, denoted by [1], of those who desire the norm N_1 as a general precept; groups [2] and [3], defined analogously; the group, denoted by [1,2] to which a person belongs if, while rejecting N_3, he is satisfied by either of the norms N_1 and N_2 as a general regulation for everyone; groups [1,3] and [2,3], defined in an analogous way; and the group [1,2,3] of those who are satisfied to make any one of the three norms the general precept.

A much more refined description of the attitudes toward the conduct regulated by $\mathfrak{N}_3$ is possible if, besides the basic norms N_1, N_2, N_3, their disjunctions are available: the norm N_1 or N_2, denoted by N_{12}; similarly defined norms N_{13} and N_{23}; and a norm N_{123}, which is the disjunction of the three basic norms. Altogether we then have what will be called the *complete system of norms* derived from $\mathfrak{N}_3$,

$$\mathfrak{N}_3^*: N_1, N_2, N_3, N_{12}, N_{13}, N_{23}, N_{123}.$$

We now assume that each person is willing to accept some of the norms of the system $\mathfrak{N}_3^*$ for the regulation of α and is unwilling to accept the others. Each individual's attitude toward α can thus be described by the (nonempty) set of the accepted norms in the system $\mathfrak{N}_3^*$. Conversely, to each nonempty subset of $\mathfrak{N}_3^*$ there corresponds a possible attitude espoused by a group of people, which in special cases may however be empty. We shall use brief, self-explanatory symbols for the nonempty subsets of $\mathfrak{N}_3^*$; for example, $N(1,12,13,123)$ and $N(2,23)$ for the quadruple N_1, N_{12}, N_{13}, N_{123} and the pair N_2, N_{23}, and write, for the sake of uniformity, $N(1)$, $N(23)$, $N(123)$,...for the sets consisting of the single norms N_1, N_{23}, N_{123},...; while we denote the groups of individuals who espouse these norms by (1,12,13,123), (2,23), (1), (23), (123), respectively.

Each of the seven groups (1),..., (1,2),..., (1,2,3) is a part of (or identical with) the corresponding group [1],..., [1,2],..., [1,2,3] for the same $\mathfrak{M}$, α, $\mathfrak{N}_3$, the only difference being that the former groups presuppose the availability of the disjunctive norms derived from $\mathfrak{N}_3$. For example, a person belongs to the group (1) if N_1 is the only regulation made possible by $\mathfrak{N}_3^*$ that he accepts; and such a person certainly also approves of N_1 and no other norm if only the three norms of $\mathfrak{N}_3$ are available, whence he also belongs to the group [1]. On the other hand, some members of [1] might join (1,12) or (1,13) or (1,12,13) or (1,12,13,123) if disjunctive norms become available. They might admit disjunctions as alternatives to N_1 or even prefer them to N_1. For similar reasons, while (1,2) is a part of (or identical with) the group [1,2], the group (12) may include members of [1], [2] or [1,2].

The group (12) of those exclusively favoring the disjunctive norm N_{12} must be distinguished from the group (1,2) of those in favor of either N_1 or N_2. The former leave everyone a choice between the conduct ordered by N_1 and the conduct ordered by N_2 whereas the latter desire a uniform precept: either N_1 or N_2, for everyone. A third group to be distinguished from both (12) and (1,2) is (1) $\cup$ (2), the union of the groups (1) and (2). It consists of those who admit only N_1 and those who admit only N_2. Someone admitting only N_1 favors neither the disjunctive norm N_{12} nor the disjunction, N_1 or N_2; and someone admitting only N_{12} or admitting N_1 or N_2 does not admit N_1 alone. Far from being identical, the three groups (12), (1,2), (1) $\cup$ (2) are mutually disjoint. The differences between them correspond to those between the following three propositions concerning an individual: he desires the disjunctive norm N_1 or N_2 as a general precept for everyone; he desires either N_1 as a general precept or N_2 as a general precept; he is either one of those desiring N_1 as a general precept or one of those desiring N_2. More generally, the desire for a disjunction must be distinguished from the disjunction of desires.

Under the assumption that $\mathfrak{N}_3^*$ includes all possible regulations of α the norm N_{123} does not formulate *any* precept and thus plays the role of the norm N_0 mentioned above. The members of the group (123) will be said to be *libertarian* with respect to α. This attitude must be distinguished from what might be called the *norm indifference* of the members of the group (1,2,3). The latter, while not insisting on a particular precept, desire some uniform regulation whereas the members of (123) reject all categorical norms. The members of the group (1,2,3,12,13,23,123) will be said to be *totally indifferent* with respect to α. Their attitude differs both from the libertarianism of the group (123) and from the mere norm indifference of those belonging to (1,2,3), who demand some uniform regulation though not insisting on a particular categorical norm.

All in all, it can be said that the admission of disjunctive norms, while not increasing the membership of any of the seven groups relative to $\mathfrak{N}_3$, in general decreases the membership of at least some of them.

There are 2^7-1 (that is, 127) nonempty subsets of $\mathfrak{N}_3^*$. But the number of *significant* attitudes is much smaller, whence the number of the (at least potentially) nonempty groups of people in relation to $\mathfrak{N}_3^*$ can be substantially reduced on the basis of some simple assumptions. Consider, for example, a member, M, of the group (1,123), that is, a person who consents only to N_1 and N_{123}. What can be the reason for M's rejection of the norm N_{12}, which prohibits behavior according to N_3 while allowing a choice between N_1 and N_2? It cannot be the prohibition of behavior according to N_3, since M admits N_1 which forbids such behavior; nor can it be the possibility of a choice between N_1 and N_2, since M admits N_{123} which includes such a choice. The only possible reason for M's rejection of N_{12} is unwillingness to grant N_2 the preference over N_3 implied by an admission of N_{12} combined with a rejection of N_{13}. But this consideration cannot be of great importance to M since he is willing to accept N_1 and thereby to exclude behavior according to N_3 completely. It thus appears that a person admitting N_1 and N_{123} could easily be persuaded to admit also N_{12}; and the same holds for N_{13}. For these reasons we will assume (without claiming to formulate a logical law or even a universal regularity) that a man admitting N_1 and N_{123} also admits N_{12} and N_{13}. This assumption implies that the groups

(+) (1,123), (1,12,123), (1,13,123)

are empty. The smallest group admitting N_1 and N_{123} that may be nonempty is (1,12,13,123). Whether or not in a concrete situation this group is actually nonempty has to be examined in each case.

Also the groups (2,123), (2,12,123) and so on are empty. More generally, let N' and N'' be two norms such that the second disjunction has the first as one of its alternatives. An example is the pair N_1 and N_{123}. Then any disjunction having N' as an alternative and being an alternative of N'' is said to be *between* N' and N''. Between N_1 and N_{123} is N_{12} as well as N_{13}. Underlying the assumption mentioned above is the following general

Principle of Intermediacy. A person accepting two norms also accepts all (if any) norms between them.

This assumption is reflected in the emptiness of groups such as (+). The applicability of the principle increases if $\mathfrak{N}$ includes more than three basic norms, say, if N is a complete system $\mathfrak{N}_4^*$, derived from four categorical norms, N_1, N_2, N_3, N_4. Then the smallest group admitting, for example, N_2 and N_{1234} according to the principle is

(2,12,23,24,123,124,234,1234),

while groups such as (2,24,1234) are empty.

Two other principles concern the disjunction and what will be called the conjunction of norms. The disjunction of N_1 and N_2 is N_{12}; that of N_{123} and N_{124} is N_{1234}. By the conjunction of N_{123} and N_{124} we mean N_{12}; the conjunction of N_{12} and N_{13} is N_1; and N_1 and N_2 are said to have no conjunction. The two principles claim that *if a person admits two norms he also admits their disjunction and their conjunction.*

For many purposes, however, the principle of conjunction must be restricted. It is possible that a person admits N_{12} and N_{13} (because he wishes everyone to be permitted to regulate his behavior according to N_1) and yet rejects the norm N_1 as categorical or as a uniform command; because in other words, he wants behavior according to N_1 to be optional. On the other hand, a person admitting N_{123} and N_{124} is very likely to admit also N_{12}, which is not categorical. Hence the following

Restricted Principle of Conjunction. The conjunction of two admitted norms is admitted at least in all cases where this conjunction is noncategorical.

About the admission of N_1 by a person admitting N_{12} and N_{13} this restricted principle makes no assertion at all. In the case of three basic norms N_1, N_2, N_3 and their disjunctions, the *Restricted Principle of Conjunction* need not be assumed since it is necessarily satisfied.

Less important is the following analogous

Restricted Principle of Disjunction. The disjunction of two admitted norms is admitted at least in all cases where one or both of the two admitted norms are noncategorical.

About the admission of N_{12} by a person admitting N_1 and N_2 this restricted principle makes no assertion at all. Nor does it rule out mere norm indifference whereas according to the (unrestricted) *Principle of Disjunction* the group (1,2,3) is empty.

A set of norms will be said to be *balanced* if, for any two norms belonging to the set, the set includes (1) all (if any) norms between them; (2) their disjunction; (3) their conjunction at least in all cases where the latter is noncategorical. This will also be expressed by saying that balanced sets are closed with respect to intermediacy, disjunction and restricted conjunction. We speak of a *completely balanced* or an *almost balanced* set of norms if the set is closed with respect to intermediacy, and to disjunction and conjunction both in unrestricted or both in restricted form, respectively.

As one readily verifies, the balanced subsets of $\mathfrak{N}_3^*$ are:

(I) complete indifference; libertarianism, that is, $N(123)$; furthermore

$$N(1),\quad N(12),\quad N(1,12),\quad N(1,2,12),\quad N(12,123),$$
$$N(1,12,13,123);$$

and the analogues of the last six sets with the roles of N_1 and N_2 and N_3 permuted; those of $N(1,2,12)$ are $N(1,3,13)$ and $N(2,3,23)$; five of the six sets have two such analogues; only $N(1,12)$ has five, namely $N(2,12)$, $N(1,13)$, $N(3,13)$, $N(2,23)$, $N(3,23)$ – altogether there are twenty-three sets which are completely balanced,

(II) the set $N(12,13,23,123)$; furthermore

$$N(1,2,12,13,23,123),\quad N(1,12,13,23,123),\quad N(12,13,123)$$

and two analogues of each of the last three sets, those of the last being $N(12,23,123)$ and $N(13,23,123)$ – altogether ten balanced but not completely balanced sets.

Almost balanced but not balanced are the four sets

$$N(1,2),\quad \mathrm{N}(1,3),\quad N(2,3),\quad N(1,2,3).$$

Of particular significance for the regulation of any behavior, β, regardless of the total number of norms proposed, are the following five sets of norms which will be said to be *fundamental* with respect to β:

$\mathfrak{A}$, the set of all norms *permitting* β;

$\mathfrak{A}'$, the set of all the other norms, that is, all norms *not permitting* β;

$\mathfrak{B}$, the set consisting of the single norm *commanding* β;

$\mathfrak{B}'$, the set of all other norms, that is, of all norms *not commanding* β;

$\mathfrak{C}$, the set of all norms making β *optional*, that is, permitted but not commanded.

Clearly, $\mathfrak{C}$ is the common part of the sets $\mathfrak{A}$ and $\mathfrak{B}'$; in symbols, $\mathfrak{C}=\mathfrak{A} \cap \mathfrak{B}'$. Also $\mathfrak{B}$ will be shown below to be the common part of other sets that are fundamental (though not with respect to β).

Given the system $\mathfrak{N}_3^*$, the five fundamental sets with respect to the behavior regulated by N_1 are:

$$\mathfrak{A}_1 = N(1,12,13,123), \quad \mathfrak{A}_1' = N(2,3,23),$$
$$\mathfrak{B}_1 = N(1), \quad \mathfrak{B}_1' = N(2,3,12,13,23,123), \quad \mathfrak{C}_1 = N(12,13,123).$$

For each of the norms N_2, N_3, there are five analogous sets.

As one easily verifies, each balanced set of norms belonging to $\mathfrak{N}_3^*$, with one single exception, is either one of the fifteen fundamental sets $\mathfrak{A}_1, \mathfrak{A}_1', \ldots, \mathfrak{B}_3, \mathfrak{B}_3', \mathfrak{C}_3$ or the common part of two or more of these fundamental sets. Combining both cases we say, briefly, that each balanced set is the common part of *some* fundamental sets, the completely balanced sets being the common parts of some of the six sets, $\mathfrak{A}_1, \mathfrak{A}_1', \mathfrak{A}_2, \mathfrak{A}_2', \mathfrak{A}_3, \mathfrak{A}_3'$.

(I) One finds $N(1)=\mathfrak{A}_2' \cap \mathfrak{A}_3'$ since N_1 is the only norm common to $\mathfrak{A}_2'=N(1,3,13)$ and $\mathfrak{A}_3'=N(1,2,12)$. Similarly,

$$N(12) = \mathfrak{A}_1 \cap \mathfrak{A}_2 \cap \mathfrak{A}_3', \quad N(1,12) = \mathfrak{A}_1 \cap \mathfrak{A}_3',$$
$$N(1,2,12) = \mathfrak{A}_3';$$
$$N(12,123) = \mathfrak{A}_1 \cap \mathfrak{A}_2 \quad \text{and} \quad N(1,12,13,123) = \mathfrak{A}_1.$$

(II) One finds $N(12,13,23,123)=\mathfrak{B}_1' \cap \mathfrak{B}_2' \cap \mathfrak{B}_3'$ and the last three balanced sets listed above are equal to $\mathfrak{B}_3'$, $\mathfrak{B}_2' \cap \mathfrak{B}_3'$, and $\mathfrak{C}_1$, respectively. The single exceptional balanced set is the total set, $\mathfrak{N}_3^*$, characterizing total indifference, which obviously is not a part of any fundamental set since it contains all of them as proper parts.

Conversely, the fundamental sets as well as the common parts of two (and hence of any) of them are easily seen to be balanced; the sets $\mathfrak{A}$, $\mathfrak{A}'$ and their common parts are completely balanced.

Since $\mathfrak{B}_1 = N(1) = \mathfrak{A}'_2 \cap \mathfrak{A}'_3 = \mathfrak{A}_1 \cap \mathfrak{A}'_2 \cap \mathfrak{A}'_3$ it follows that (in accordance with a remark above) each of the sets $\mathfrak{B}$ is the common part of other fundamental sets, as is each set $\mathfrak{C}$. Thus *each balanced set, except the total set* $\mathfrak{N}^*_3$, *is the common part of some of the nine fundamental sets,*

$$\mathfrak{A}_1, \mathfrak{A}_2, \mathfrak{A}_3, \mathfrak{A}'_1, \mathfrak{A}'_2, \mathfrak{A}'_3, \mathfrak{B}'_1, \mathfrak{B}'_2, \mathfrak{B}'_3,$$

the completely balanced sets being the common parts of some of the first six of them.

Using a simple theorem concerning finite sets one can extend this characterization of balanced sets from three norms to k norms, $N_1, N_2, \ldots, N_k$, and the total system, $\mathfrak{N}^*_k$, of disjunctive norms: *A proper subset of* $\mathfrak{N}^*_k$ *is balanced if and only if it is the common part of some of the 3k fundamental sets*

$$\mathfrak{A}_1, \mathfrak{A}_2, \ldots, \mathfrak{A}_k, \mathfrak{A}'_1, \mathfrak{A}'_2, \ldots, \mathfrak{A}'_k, \mathfrak{B}'_1, \mathfrak{B}'_2, \ldots, \mathfrak{B}'_k,$$

the completely balanced sets being the common parts of some of the first 2k of these sets. $\mathfrak{N}^*_k$ *itself is completely balanced.*

Some important relations between the $3k$ fundamental sets lend themselves to graphical representation. Figures 1 and 2 illustrate relations involving N_1 and N_2, that is, between the ten sets

(†) $\quad \mathfrak{A}_1, \mathfrak{A}'_1, \mathfrak{B}_1, \mathfrak{B}'_1, \mathfrak{C}_1, \mathfrak{A}_2, \mathfrak{A}'_2, \mathfrak{B}_2, \mathfrak{B}'_2, \mathfrak{C}_2;$

but any two of k norms might have been chosen. The relations are valid regardless of the total number, k, of norms in $\mathfrak{N}$. In both diagrams, each of the sets (†) is represented by a point.

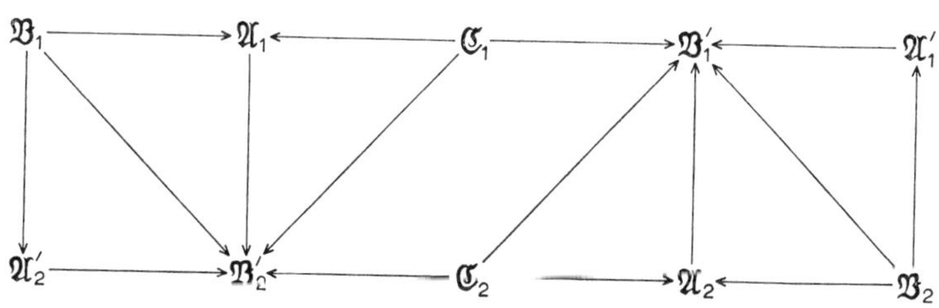

Fig. 1.

Figure 1 illustrates the subset relation. An arrow is drawn from one point to another if the set represented by the first point is a subset of the set represented by the second. For example, three arrows issue from $\mathfrak{B}_1$

(pointing to $\mathfrak{A}_1$, to $\mathfrak{A}'_2$, and to $\mathfrak{B}'_2$); and five arrows point to $\mathfrak{B}'_2$ (issuing from $\mathfrak{B}_1$, $\mathfrak{A}_1$, $\mathfrak{C}_1$, $\mathfrak{A}'_2$, and $\mathfrak{C}'_2$). Indeed, $\mathfrak{B}_1$ is obviously a subset of $\mathfrak{A}_1$, $\mathfrak{A}'_2$ and $\mathfrak{B}'_2$; and $\mathfrak{B}'_2$ has $\mathfrak{B}_1$, $\mathfrak{A}_1$, $\mathfrak{C}_1$, $\mathfrak{A}'_2$, and $\mathfrak{C}_2$ among its subsets. If a person's attitude corresponds to the end point of an arrow, then he accepts all norms accepted by those whose attitude corresponds to the initial point.

In Figure 2, the dotted lines indicate the disjointness of the sets joined by a line and the incompatibility of the attitudes represented by those points. For example, the set $\mathfrak{C}_1$ is disjoint from $\mathfrak{A}'_1$, $\mathfrak{B}_1$, and $\mathfrak{B}_2$.

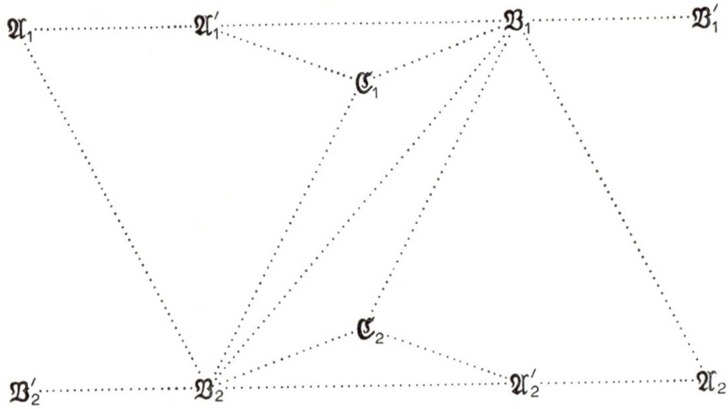

Fig. 2.

It is easy to verify that if k exceeds 2, that is, if $\mathfrak{N}$ includes three or more norms, then these diagrams represent *all* relations of inclusion and disjointness between the ten sets (†) and all relations of implication and incompatibility between the positions they symbolize. Only in the case $k=2$, where $\mathfrak{N}$ consists of the norms N_1 and N_2 alone, do there exist further relations, because in this case the ten sets (†) collapse into five:

$$\mathfrak{A}_1 = N(1,12) = \mathfrak{B}'_2, \quad \mathfrak{A}_2 = N(2,12) = \mathfrak{B}'_1,$$
$$\mathfrak{A}'_1 = N(2) = \mathfrak{B}_2, \quad \mathfrak{A}'_2 = N(1) = \mathfrak{B}_1, \quad \mathfrak{C}_1 = N(12) = \mathfrak{C}_2\,;$$

and in Figures 1 and 2, the upper and lower lines coincide. Disjunctive norms permit more subtle distinctions of attitude than the mere approval, disapproval, and indifference available in presence of only categorical norms. Among various possible interpretations of these three attitudes we choose the following. By approving or disapproving a norm N one accepts only regulations which, respectively, do or do not permit behavior according to N, while indifference is manifested by the acceptance of any other regulation. Thus in the case of the complete system $\mathfrak{N}_3^*$ derived from three norms, N_1, N_2, N_3, the groups $\mathfrak{M}_1^+$ and $\mathfrak{M}_1^-$ (of persons approving or disapproving of N_1, respectively), consist of those who accept any balanced subset of $\mathfrak{A}_1$ and $\mathfrak{A}_1'$, respectively, while the group $\mathfrak{M}_1^0$, made up of those who are indifferent to N_1, includes all persons with other balanced attitudes. Thus

$$\begin{aligned}\mathfrak{M}_1^+ &= (1) \cup (12) \cup (13) \cup (123) \cup (1{,}12) \cup (1{,}13) \cup \\ &\quad \cup (12{,}123) \cup (13{,}123) \cup (12{,}13{,}123) \cup (1{,}12{,}13{,}123), \\ \mathfrak{M}_1^- &= (2) \cup (3) \cup (23) \cup (2{,}23) \cup (3{,}23) \cup (2{,}3{,}23)\,,\end{aligned}$$

and $\mathfrak{M}_1^0$ is the union of the adherents to the thirty-three minus ten minus six, that is, seventeen other balanced groups. There are twenty-seven groups of consentience,

$$\mathfrak{M}_1^+ \cap \mathfrak{M}_2^+ \cap \mathfrak{M}_3^+\,,\ \ \mathfrak{M}_1^+ \cap \mathfrak{M}_2^+ \cap \mathfrak{M}_3^0, \ldots, \mathfrak{M}_1^- \cap \mathfrak{M}_2^- \cap \mathfrak{M}_3^-\,.$$

The last group as well as the three groups with one indifferent and two negative attitudes are empty. For in the case of three categorical norms, disapproval of two of them (being interpreted as the rejection of the two as well as of all disjunctions admitting them as alternatives) entails the approval of the third. The twenty-three other groups of consentience are seen to include

$$\begin{aligned}\mathfrak{M}_1^+ \cap \mathfrak{M}_2^+ \cap \mathfrak{M}_3^+ &= (123)\,, \\ \mathfrak{M}_1^+ \cap \mathfrak{M}_2^+ \cap \mathfrak{M}_3^0 &= (12{,}123)\,, \\ \mathfrak{M}_1^+ \cap \mathfrak{M}_2^+ \cap \mathfrak{M}_3^- &= (12)\,, \\ \mathfrak{M}_1^+ \cap \mathfrak{M}_2^0 \cap \mathfrak{M}_3^0 &= (12{,}13{,}123) \cup (1{,}12{,}13{,}123)\,, \\ \mathfrak{M}_1^+ \cap \mathfrak{M}_2^- \cap \mathfrak{M}_3^- &= (1)\,, \\ \mathfrak{M}_1^- \cap \mathfrak{M}_2^0 \cap \mathfrak{M}_3^0 &= (2{,}3{,}23)\,, \\ \mathfrak{M}_1^+ \cap \mathfrak{M}_2^0 \cap \mathfrak{M}_3^- &= (1{,}12)\,.\end{aligned}$$

The last group has five analogues in which the roles of N_1, N_2, N_3 are permuted; each of the five preceding groups has two such analogues. These twenty-two groups of consentience obviously contain twenty-five of the thirty-three groups with balanced attitudes. The twenty-third group of consentience, that is, $M_1^0 \cap M_2^0 \cap M_3^0$, is the union of the remaining eight; if, for the sake of brevity, the array 12, 13, 23, 123 be replaced by three dots, then $M_1^0 \cap M_2^0 \cap M_3^0$ is

$$(\ldots) \cup (1, \ldots) \cup (2, \ldots) \cup (3, \ldots) \cup (1,2, \ldots) \cup (1,3, \ldots) \cup (2,3, \ldots) \cup (1,2,3, \ldots).$$

In the establishment of groups of compatibility, logical relations between the proposed basic norms may be important. But if a norm is a disjunction of two precepts that are not among the basic norms, then it may be treated as a categorical norm; for example, the norm to further one's scientific or artistic education may be considered as categorical in the absence of a norm concerning scientific education and one concerning artistic education. On the other hand, if the system $\mathfrak{N}$ proposed for the regulation of α is an $\mathfrak{N}_7$ consisting of the seven norms belonging to $\mathfrak{N}_3^*$, then $\mathfrak{N}_7^*$ is identical with $\mathfrak{N}_3^*$.

It is also possible that $\mathfrak{N}$ consists of categorical norms and some, but not all, of their disjunctions. An important example is the case where $\mathfrak{N}$ consists of categorical norms and the proposal, N_0, to leave α unregulated. If $k=3$ then, besides the seven groups [1], [2],..., [1,2,3] mentioned on p. 53, eight further groups appear:

$$[0], \quad [0,1], \quad [0,2], \quad [0,3], \quad [0,1,2], \quad [0,1,3], \quad [0,2,3], \quad [0,1,2,3].$$

The members of the last group may be said to be *totally indifferent* to α, while those of [0] may be called *libertarian* with respect to α. (No disjunctive norms are here being considered. As before, [1,2] consists of those who wish that either N_1 be a general rule or N_2; and, similarly, the members of [0,1] desire N_1 or nonregulation.)

One may again interpret approval and disapproval of N_1 as the exclusive acceptance of regulations that respectively do and do not permit behavior according to N_1; and indifference as accepting some regulations which do, and some which do not, permit such behavior. Since the latter is compatible with nonregulation this interpretation leads to the following triple of formulae

$$\mathfrak{M}_1^+ = [1] \cup [0,1] \cup [0], \qquad \mathfrak{M}_1^- = [2] \cup [3] \cup [2,3],$$
$$\mathfrak{M}_1^0 = [1,2] \cup [1,3] \cup [1,2,3] \cup [0,2] \cup [0,3] \cup [0'],$$

where

$$[0'] = [0,1,2] \cup [0,1,3] \cup [0,2,3] \cup [0,1,2,3],$$

and to two analogous triples for N_2 and N_3. Of the 27 groups of consentience, all are empty except possibly the following 11 groups

$$\begin{aligned} &\mathfrak{M}_1^+ \cap \mathfrak{M}_2^- \cap \mathfrak{M}_3^- = [1], \\ &\mathfrak{M}_1^0 \cap \mathfrak{M}_2^0 \cap \mathfrak{M}_3^- = [1,2], \quad \mathfrak{M}_1^+ \cap \mathfrak{M}_2^0 \cap \mathfrak{M}_3^0 = [0,1], \\ &\mathfrak{M}_1^+ \cap \mathfrak{M}_2^+ \cap \mathfrak{M}_3^+ = [0], \quad \mathfrak{M}_1^0 \cap \mathfrak{M}_2^0 \cap \mathfrak{M}_3^0 = [1,2,3] \cup [0'], \end{aligned}$$

where each of the first three formulae has two analogues.

Another interpretation may take approval of N as the demand for behavior according to N; disapproval as the exclusive acceptance of regulations that make such behavior unenforceable or, in other words, rejection of the possibility that N be a general command; indifference as any other attitude. One then finds

$$\begin{aligned} &\mathfrak{M}_1^+ = [1], \quad \mathfrak{M}_1^- = [2] \cup [3] \cup [2,3] \cup [0] \cup [0,2] \cup [0,3] \cup [0,2,3], \\ &\mathfrak{M}_1^0 = [0,1] \cup [1,2] \cup [1,3] \cup [0,1,2] \cup [0,1,3] \cup [1,2,3] \cup [0,1,2,3]. \end{aligned}$$

Again at most 11 groups of consentience are (potentially) nonempty; namely, the following five together with the six analogues of the first three:

$$\begin{aligned} &\mathfrak{M}_1^+ \cap \mathfrak{M}_2^- \cap \mathfrak{M}_3^- = [1], \\ &\mathfrak{M}_1^0 \cap \mathfrak{M}_2^0 \cap \mathfrak{M}_3^- = [1,2] \cup [0,1,2], \\ &\mathfrak{M}_1^0 \cap \mathfrak{M}_2^- \cap \mathfrak{M}_3^- = [0,1], \\ &\mathfrak{M}_1^- \cap \mathfrak{M}_2^- \cap \mathfrak{M}_3^- = [0], \\ &\mathfrak{M}_1^0 \cap \mathfrak{M}_2^0 \cap \mathfrak{M}_3^0 = [1,2,3] \cup [0,1,2,3]. \end{aligned}$$

If approval of N is interpreted as the demand for behavior according to N, disapproval as the demand for a regulation not permitting such behavior, and indifference as any other attitude, then

$$\mathfrak{M}_1^+ = [1], \quad \mathfrak{M}_1^- = [2] \cup [3] \cup [2,3],$$

while M_1^0 is the union of the 11 other groups. In this case, only seven groups can be nonempty, namely, the following two with their four analogues:

$$\mathfrak{M}_1^+ \cap \mathfrak{M}_2^- \cap \mathfrak{M}_3^- = [1], \quad \mathfrak{M}_1^0 \cap \mathfrak{M}_2^0 \cap \mathfrak{M}_3^- = [1,2],$$

and $M_1^0 \cap M_2^0 \cap M_3^0$, which is the union of [1,2,3] and the eight groups admitting nonregulation.

If the demand for nonregulation be denoted by N_0, then in none of these interpretations does N_0 play a role analogous to that of N_1, N_2, N_3.

It is clear that all preceding remarks can be extended from 3 to any natural number k of norms, $N_1, N_2, \ldots, N_k$. For example, let approval and disapproval of a norm be interpreted as desires that the prescribed behavior be, respectively, commanded and unenforceable. Then each of the $3k$ groups of attitudes toward one of the k norms, $\mathfrak{M}_1^+, \mathfrak{M}_1^-, \ldots, \mathfrak{M}_k^-, \mathfrak{M}_k^0$, is the union of some of the $2^{k+1}-1$ groups [0], [1],..., [0,1,..., k] corresponding to the nonempty sets of numbers 0, 1,..., k; and so is each of the potentially nonempty groups of consentience. Conversely, however, only $2k+1$ of the $2^{k+1}-1$ groups are determined by the groups of consentience, namely,

$$[0] = \mathfrak{M}_1^0 \cap \mathfrak{M}_2^0 \cap \ldots \cap \mathfrak{M}_k^0,$$
$$[1] = \mathfrak{M}_1^+ \cap \mathfrak{M}_2^- \cap \ldots \cap \mathfrak{M}_k^-,$$
$$[0,1] = \mathfrak{M}_1^0 \cap \mathfrak{M}_2^- \cap \ldots \cap \mathfrak{M}_k^-,$$

and the $2(k-1)$ analogues of the last two groups. But for any h exceeding 1

$$[1,2, \ldots, h] \cup [0,1,2, \ldots, k] = \mathfrak{M}_1^0 \cap \mathfrak{M}_2^0 \cap \ldots \cap \mathfrak{M}_h^0 \cap \mathfrak{M}_{h+1}^- \cap \mathfrak{M}_{h+2}^- \cap \ldots \cap \mathfrak{M}_k^- .$$

The two terms on the left side, separately, are not determined by the groups of consentience since, for example, the members of [1,2] and of [0,1,2] have the same attitudes toward each of the k norms, namely, indifference to N_1 and N_2, and disapproval of $N_3, N_4, \ldots, N_k$.

In order to determine the $3k$ groups $\mathfrak{M}_1^+, \ldots, \mathfrak{M}_k^0$ one may present a list of all k proposed regulations of α to each member of the class $\mathfrak{M}$ and ask him which of the three possible attitudes (approval, disapproval, or indifference) he takes toward each. In order to determine the $2^{k+1}-1$ groups [0], [1],..., [0,1,..., k] one must ask which set of norms (including nonregulation) he desires. The remark of the preceding paragraph demonstrates that the knowledge of a person's answer to the second question makes it possible to conclude his answer to the first, whereas from his answer to the first question it is not possible to infer his complete answer to the second.

4. A PERSON'S DEMANDS ON HIMSELF AND ON OTHERS

It often happens that someone maintains a certain attitude concerning a norm himself but does not attach importance to the attitudes of others; he may even desire that they maintain a position different from his own. In the latter case, he is not compatible with persons sharing his views and prefers not to be grouped with them. His group of consentience is not a suitable environment for him. A group will be called a *compatibility group* if its members are mutually compatible.

The significance of these matters becomes clear when one recalls that what the categorical imperative, at least in its externalized form, requires is precisely the agreement between norms that a person is to follow himself and norms that everyone else is to follow. This situation will now be examined more closely.

As a simple example, consider the norm that requires certain forms of courtesy. Persons who pattern their behavior in accordance with this norm will be said to be *polite*; violators of the norm will be called *impolite*. Another partition divides people into those who want to associate only with polite persons and those to whom the courtesy of the people with whom they associate is a matter of indifference. For the sake of a concise terminology, we shall refer to the former as *sensitive*, to the latter as *insensitive*.

The two partitions into polite and impolite, sensitive and insensitive persons overlap. Thus, four categories of persons result:

(*a*) polite, sensitive persons,
(*b*) polite, insensitive persons,
(*c*) impolite, sensitive persons,
(*d*) impolite, insensitive persons.

In a particular class of people, one or more of these four categories may of course be empty; for example, the class may not include impolite, sensitive persons.

Abstracting from all other conditions, what can be said about the relationships between people of these four categories? The following is clear:

(1) Any member of category (*a*) – briefly of (*a*) – is willing to associate with members of (*a*) and (*b*), but not with members of (*c*) and (*d*).

(2) Any member of (*b*) is willing to associate with any member of (*a*), (*b*), (*c*), (*d*).

(3) Any member of (*c*) is willing to associate with members of (*a*) and (*b*), but neither with other members of (*c*) nor with members of (*d*).

(4) Any member of (*d*) is willing to associate with anyone in (*a*), (*b*), (*c*), (*d*).

In order for association to take place between two persons, *both* must be willing. This mutual willingness occurs in the following cases:

(I) A member of (*a*) can associate with every other member of (*a*) and with every member of (*b*), but not with members of (*c*) or (*d*).

(II) A member of (*b*) can associate with everyone.

(III) A member of (*c*) can associate with any member of (*b*), but not with members of (*a*) or (*d*) nor with other members of (*c*).

(IV) A member of (*d*) can associate with every other member of (*d*) as well as with every member of (*b*).

The preceding remarks can be represented both in matrix and in diagram form. In the matrix form, each point of a rectangular scheme is contained in two linear arrays: a vertical column and a horizontal row. Each of these arrays corresponds to one of the four categories (*a*)–(*d*).

The first matrix illustrates the statements (1)–(4). In column (*a*) an asterisk is placed in those rows that correspond to categories with whose members the members of category (*a*) are willing to associate; that is, in the rows (*a*) and (*b*). In column (*b*) an asterisk is placed in every row, since each member of category (*b*) is willing to associate with every member of every category. The asterisks in columns (*c*) and (*d*) are placed in the same rows as those in columns (*a*) and (*b*), respectively.

Results (I)–(IV) are summarized in the second matrix

	a	*b*	*c*	*d*
a	*	*	*	*
b	*	*	*	*
c		*		*
d		*		*

	a	*b*	*c*	*d*
a	*u*	*u*		
b	*u*	*u*	*u*	*u*
c		*u*		
d		*u*		*u*

Here, *u* (for union) is placed at the intersection of a column and a row if and only if all members of the column category can associate with all members of the row category. In contrast to the first matrix the second is symmetrical with respect to the main diagonal running from top left to bottom right; for if association is possible between one person and a second, then it is also possible between the second and the first.

The second matrix can be derived from the first. Place *u* at a point of the second matrix if and only if the first contains an asterisk at the corresponding point as well as at the point symmetrical with respect to the main diagonal. (The first matrix can of course not be derived from the second.)

Figure 3, which illustrates these relationships shows the letters *a, b, c, d.* A letter is encircled if any two members of the corresponding category are willing to associate with one another. In Figure 3a, corresponding to the first matrix, an arrow is drawn from a letter, *x*, to a letter, *y*, if and only if any member of the category (*x*) is willing to associate with any member of the category (*y*).

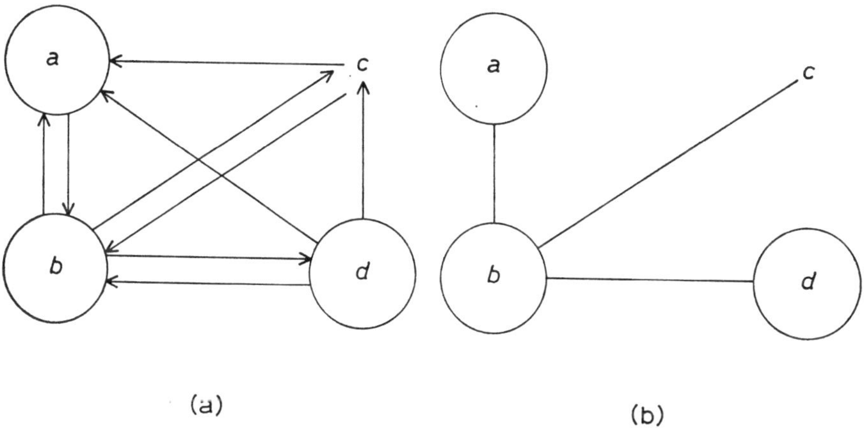

Fig. 3.

Figure 4b, corresponding to the second scheme, is obtained by connecting two points with a line if and only if in Figure 3a they are joined by double arrows. The points connected by lines in this diagram correspond to the categories between which association is possible.

Simple reflection shows that one example of a partition into compatibility groups may be obtained in the following manner.

(1) All persons in category (*a*) are included in one group.

(2) All persons in (*d*) are included in another group.

(3) For every single member of (*c*) a separate group is formed consisting of that individual alone. (No two members of category (*c*) can be

included in any one group since they do not want to associate with one another.)

If z is the number of persons in (*c*) then $2+z$ groups have been formed at this stage.

(4) Each member of (*b*) is assigned to one of these $2+z$ groups. They may all be assigned to one of the groups or distributed over several of them.

Further reflection leads to a survey of *all possible partitions* of a class of people into (jointly exhaustive and mutually disjoint) compatibility groups. While each group of the partition includes *only* mutually associable persons it is not required, and in many cases not possible, that each group include *all* individuals who can associate with one of its members. Every such partition can be obtained in the following manner.

(1) Category (*a*) is partitioned into a number, say n_1, of groups to each of which some of the members of (*b*) are adjoined. ('Some' here and in the sequel includes the possibilities of *none* and of *all*.)

(2) Category (*d*) is partitioned into a number, say n_2, of groups to each of which some of the members of (*b*) are adjoined.

(3) If z is the number of members of (*c*), then z groups are formed each of which consists of one member of (*c*) and some members of (*b*).

(4) All members of (*b*) who have not been assigned to any of the n_1+n_2+z groups already formed are partitioned into a number, say n_3, of groups. If the entire category (*b*) has been distributed among the n_1+n_2+z first groups, then of course $n_3=0$.

If the total class actually includes persons belonging to categories (*a*) and (*d*), then each partition of the class into groups of mutually associable persons must consist of at least $2+z$ groups. This minimum number of groups is attained in the example given above, where $n_1=1$, $n_2=1$, $n_3=0$; that is, where all members of (*a*) are assigned to a single group, and similarly all members of (*d*), while all members of (*b*) are assigned to one of the $2+z$ first groups. (It should be clear that the arbitrarily chosen numbers n_1, n_2, n_3 do by no means determine a partition. There are, for example, many ways of partitioning (*a*) into a given number, n_1, of groups depending upon the numbers of individuals assigned to those groups.)

Various objectives may motivate the partitioning of a class. One is to form groups of roughly equal size. With this aim in mind one will use the members of category (*b*) to equalize the groups obtained. One of many other possibilities is to place the members of (*b*) in the other groups in numbers roughly proportional to the size of those groups.

The simple example of the courtesy norm has many formal analogues such as the norms requiring one to be helpful or to pay attention to external appearance. More important, it makes us aware of some noteworthy general characteristics of groups of people.

Compatibility is not in general a transitive relation. In the example, members of category (*a*) and members of (*c*) are compatible with members of (*b*) without being compatible with one another. We shall call a person *self-compatible* if he is compatible with every person assuming toward all relevant norms the same attitude that he himself assumes. Naturally, this concept may be relativized to special norms or classes of norms. In the preceding example the members of categories (*a*), (*b*) and (*d*) are self-compatible; those of (*c*) are self-incompatible.

Suppose that a class of people includes a group (i) of consentient but self-incompatible individuals and assume first that all other members of the class have already been partitioned into compatibility groups. If the number of the members of (i) does not exceed the number of all of the groups of the partition, then one can distribute (i) by assigning its members to separate groups. Otherwise, there must be compatibility groups each consisting solely of a single member of (i); in other words, some members of the group (i) can only exist in isolation unless they change their attitude. If, however, the partition of the total class begins with the formation of compatibility groups some of which include one single member of the group (i), then it may be possible to accommodate a larger number or perhaps all of the self-incompatible individuals in groups, if only in small groups. A population partitioned into partriarchically organized families is an example.

In the case of the courtesy norm the members of categories (*b*) and (*d*) are willing to be associated with everyone or, as we shall say, are *all-accepting*. The members of categories (*a*) and (*b*) have the property that everyone is willing to associate with them and may be called *all-accepted*. A person can associate with everyone or is, as we may say, *all-compatible*, if and only if he is both all-accepting and all-accepted. Here an important observation must be made. The characteristics of being all-accepting, all-accepted and all-compatible can be understood either in an *absolute* sense; that is, with respect to persons of all *conceivable* attitudes toward the norms under consideration; or they may be understood *relative* to some class of people which may not include representatives of all attitudes. In the example of courtesy, the members of (*b*) are all-compatible and they are the only individuals who are all-compatible provided that there actually exist members of the other categories. If however, the categories (*c*) and (*d*) are empty, then also the members of (*a*) are all-compatible.

If the all-compatible individuals of a class are ignored until all other members have been distributed into compatibility groups, then the all-compatible people may be assigned to the constituted groups in any way whatever or to groups by themselves to complete the partition of the class. In the example of courtesy, the (all-compatible) individuals of category (*b*) have been assigned in this way.

We shall call a person *simple* (*schlicht*) if he wishes that the people around him assume the same attitude toward norms that he himself assumes – a concept that may of course be restricted to a particular system of norms or to a single norm. Polite, sensitive persons are simple; impolite, sensitive persons are not. In a certain sense one can say that the simple persons arc just those who wish to follow the categorical imperative.

Clearly, each simple person is self-compatible, whereas a self-compatible person is not necessarily simple. Members of category (*a*) are self-compatible without being simple. Two simple individuals are not necessarily compatible. A polite, sensitive person and an impolite person who wants to associate only with people of his own kind are both simple and yet incompatible.[1] Everyone's obeying the categorical imperative, therefore, is not a guarantee for harmony. On the other hand, impolite sensitive and polite insensitive persons are compatible without obeying the categorical imperative. Hence the following result.

The categorical imperative is not a sufficient precept to guarantee compatibility of people, since simple persons may be incompatible. On the other hand, *the categorical imperative is not a necessary precept to eliminate modes of behavior that prevent integration into compatible groups*, since persons not obeying the imperative are not necessarily incompatible with larger groups.

The preceding considerations can be developed within an extension of the frame of the third note. There, it has been assumed that each member,

[1] This example is elaborated in Section III of the Postscript to the English edition; it is generalized in the fifth note of this chapter.

M, of $\mathfrak{M}$ accepts some norms of $\mathfrak{N}$ and does not accept others. What follows is based on the refined assumption that M accepts (1) some norms (and not others) as regulations *for his own behavior*, (2) some norms (and not others) as regulations *for the behavior of other people.*

The norms actually *admitted* by M are those which he accepts as regulations both for his own and other people's conduct.

Naturally, M accepts (1) only norms that permit at least one mode of behavior acceptable to himself, (2) only norms that prohibit all types of behavior disagreeable to him in others. In analogy to the principles concerning balanced attitudes formulated in the third note, we now assume the following

Completeness Principle. Each member of $\mathfrak{M}$ accepts (1) *all norms that permit at least one mode of behavior acceptable to him for his own conduct and* (2) *all norms that prohibit all modes of behavior disagreeable to him in others.*

Corollary. As far as M's own conduct is concerned, if M accepts a norm N, then he accepts each disjunctive norm including N as an alternative. As far as others are concerned, if M accepts N, then he accepts each norm included in N as an alternative.

For example if $\mathfrak{N}$ is $\mathfrak{N}_3^*$, and if (1) for himself,

M accepts	then he accepts	thus	that is
N_1	N_{12}, N_{13}, N_{123}	$N(1,12,13,123)$	$\mathfrak{A}_1$
N_{12}	N_{123}	$N(12,123)$	$\mathfrak{A}_1 \cap \mathfrak{A}_2$
N_1 and N_2	all norms but N_3		$\mathfrak{B}'_3$
N_1 and N_2 and N_3	all norms		$\mathfrak{N}^*_3$

if (2) for others, M accepts N_{12}, then he accepts N_1 and N_2 thus $N(1,2,12)$, that is, $\mathfrak{A}'_3$; and if M accepts N_{123}, then he accepts all norms, thus $\mathfrak{N}_3^*$.

Empirically, there are some exceptions to the *Completeness Principle.* For example, if besides two conflicting groups a class of people includes a few vacillating individuals, then it occasionally happens that a member of one of the two groups tolerates members of his own as well as of the opposing group in his environment, but none of the vacillating individuals. This behavior violates part (2) of the Corollary for the reason that the stability of others is not guaranteed by disjunctive norms.

In the third note it has been assumed that each member of $\mathfrak{M}$ admits at least one norm of $\mathfrak{N}$. In the sequel it will be assumed that everyone accepts at least one norm (though possibly only the disjunctive norm that amounts to non-regulation) for himself, and at least one norm for the conduct of others, but not that he necessarily admits any norm in both capacities.

Two persons will be said to be *completely consentient* if they accept the same norms for themselves, and the same norms for others. They will be said to be *compatible* if either consents to at least one norm for others that the other accepts for himself.

If M_1 admits no norm, and M_2 is completely consentient with M_1, then M_1 and M_2 clearly are incompatible and both are self-incompatible. If, on the other hand, M_1 does admit a norm, then so does every person, M_2, who is completely consentient with M_1; and the two are compatible while either is self-compatible. It follows that *a member of* $\mathfrak{M}$ *is self-compatible if and only if he admits at least one norm.*

As an example, let $\mathfrak{N}$ be $\mathfrak{N}_3^*$ and consider only groups of people with balanced attitudes. According to the two parts of the Corollary of the *Completeness Principle*, $\mathfrak{M}$ is divided in two ways:

(1) into 18 groups each accepting for itself one of the following groups of norms:

$$\mathfrak{A}_1, \quad {}_2, \mathfrak{A}_3, \mathfrak{A}_1 \cap \mathfrak{A}_2, \mathfrak{A}_1 \cap \mathfrak{A}_2, \mathfrak{A}_2 \cap \mathfrak{A}_3, \mathfrak{A}_1 \cap \mathfrak{A}_2 \cap \mathfrak{A}_3;$$
$$\mathfrak{C}_1, \mathfrak{C}_2, \mathfrak{C}_3, \mathfrak{B}_1' \cap \mathfrak{B}_2' \cap \mathfrak{B}_3', \mathfrak{B}_2' \cap \mathfrak{B}_3', \mathfrak{B}_1' \cap \mathfrak{B}_3', \mathfrak{B}_1' \cap \mathfrak{B}_2',$$
$$\mathfrak{B}_3', \mathfrak{B}_2', \mathfrak{B}_1', \mathfrak{N}_3^*;$$

(2) into 7 groups each accepting for others one of the following groups of norms:

$$\mathfrak{B}_1, \mathfrak{B}_2, \mathfrak{B}_3, \mathfrak{A}_1', \mathfrak{A}_2', \mathfrak{A}_3', \mathfrak{N}_3^*.$$

The first six groups of the second division aim essentially at prohibitions, as indicated by the letters $\mathfrak{A}'$ (also $\mathfrak{B}_1 = \mathfrak{A}_2' \cap \mathfrak{A}_3'$). Persons accepting $\mathfrak{N}_3^*$ for others are totally indifferent with respect to the behavior under consideration.

The two divisions overlap yielding 126 groups. If g is a group of the first division, and h one of the second, then a member of $g \cap h$ admits those, and only those, norms accepted by g as well as h. For example, a person accepting $\mathfrak{A}_1$ for himself, and $\mathfrak{N}_3^*$ for others (that is, a totally indifferent individual who wants to follow N_1) can only admit $N(1,12,13,123)$; the admission of any other regulation might

TABLE I

	B_1	B_2	B_3	A'_1	A'_2	A'_3	$N_3{}^*$
A_1	(1)	–	–	–	(1,13)	(1,12)	(1,12,13,123) = (1 ...)
A_2	–	(2)	–	(2,23)	–	(2,12)	(2,12,23,123) = (2 ...)
A_3	–	–	(3)	(3,23)	(3,13)	–	(3,13,23,123) = (3 ...)
$A_1 \cap A_2$	–	–	–	–	–	(12)	(12,123) = (12.)
$A_1 \cap A_3$	–	–	–	–	(13)	–	(13,123) = (13.)
$A_2 \cap A_3$	–	–	–	(23)	–	–	(23,123) = (23.)
$A_1 \cap A_2 \cap A_3$	–	–	–	–	–	–	(123) = (...)
C_1	(1)	(2)	–	–	(13)	(12)	(12,13,123)
C_2	(1)	–	(3)	(23)	–	(12)	(12,23,123)
C_3	–	(2)	(3)	(23)	(13)	–	(13,23,123)
$B'_1 \cap B'_2 \cap B'_3$	–	–	–	(23)	(13)	(12)	(*)
$B'_2 \cap B'_3$	(1)	–	–	(23)	(1,13)	(1,12)	(1,*)
$B'_1 \cap B'_3$	–	— (2)	–	(2,23)	(13)	(2,12)	(2,*)
$B'_1 \cap B^*_2$	–	–	(3)	(3,23)	(3,13)	(12)	(3,*)
B'_3	(1)	(2)	–	(2,23)	(1,13)	(1,2,12)	(1,2,*)
B'_2	(1)	–	(3)	(3,23)	(1,3,13)	(1,12)	(1,3,*)
B'_1	–	(2)	(3)	(2,3,23)	(3,13)	(2,12)	(2,3,*)
N^*_3	(1)	(2)	(3)	(2,3,23)	(1,3,13)	(1,2,12)	(1,2,3,*)

prevent him from living according to N_1. A person accepting $\mathfrak{A}_2$ for himself, and $\mathfrak{A}_1'$ for others (that is, an individual who wants to follow N_2, and to forbid N_1 for others) admits $N(2,23)$. A person accepting $\mathfrak{A}_1 \cap \mathfrak{A}_2$ for himself and $\mathfrak{A}_2'$ for others (that is, an individual who desires the choice to follow N_1 or N_2, but who forbids behavior according to N_2 to others) does not find any provision in $\mathfrak{N}_3^*$ that would accomodate his wishes. He is, therefore, self-incompatible as is each member of his group of total consentience. Of the 126 groups, 48 are self-incompatible.

The groups are tabulated in Table I. Instead of Gothic letters, A, B, C, N are used for the groups of norms. Each self-incompatible group is marked by a dash. That a person who accepts A_2 for himself, and A_1' for others, admits $N(2,23)$ and thus belongs to the group (2,23) is expressed by placing (2,23) in the row of A_2 and the column of A_1'.

The symbols with dots appearing in the last column of Table I are abbreviations used in Figure 5. In the last eight rows, each asterisk stands for 12,13,23,123.

Of particular significance are the groups in the first seven rows. The exclusive use of the letter A indicates that their members desire only per-

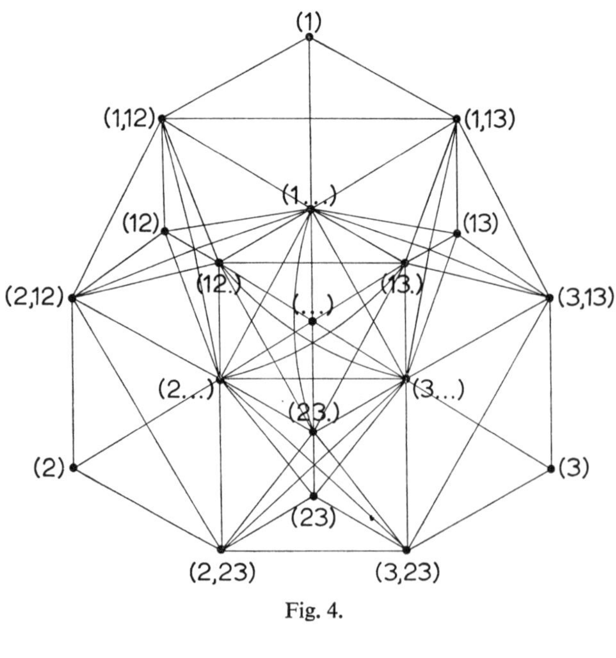

Fig. 4.

mission of certain behaviors for themselves. 19 of these 49 groups are self-compatible. Their compatibility relationships are illustrated in Figure 4.

5. Several modes of behavior

For the sake of brevity, many problems related to the subject matter of this book must be passed over in these notes: people's indecision with respect to some norms; various degrees of preference or disinclination; changes of attitudes in the course of time; divergences of actions, words, and thoughts. And mathematical tools of any degree of complication have been avoided.

Only one generalization of the theory developed in the third note will be briefly discussed. There, the frame of reference consists of a class, $\mathfrak{M}$, of people, a behavior or a situation, α, and the class $\mathfrak{N}$ of all norms for the regulation of α. In what follows, the frame includes $\mathfrak{M}$ and three modes of behavior, α, α', α'', and three classes of norms, $\mathfrak{N}$, $\mathfrak{N}'$, $\mathfrak{N}''$, consisting of k, k', k'' norms, respectively. It will be assumed that besides all regulations proposed for α, α', α'', the classes $\mathfrak{N}$, $\mathfrak{N}'$, $\mathfrak{N}''$ also include the proposals (likewise considered to be norms) that the respective modes of behavior be left unregulated.

There are $k \cdot k' \cdot k''$ ways of selecting exactly one norm from $\mathfrak{N}$, one from $\mathfrak{N}'$, and one from $\mathfrak{N}''$. But not all triples of norms thus obtained are free from contradiction; e.g., a norm concerning α and a norm concerning α' may be incompatible. A triple of compatible norms will be called a *code* with respect to α, α', α''. Each member of $\mathfrak{M}$ accepts at least one code.

By the *compatibility base* with respect to α, α', α'' of two individuals or two groups we mean the set of all codes admitted by both individuals or by all members of both groups. That base may be empty; in other words, none of the codes may be accepted by both individuals or both groups; in this case, the individuals or groups are said to be *incompatible*. By a *compatibility group* with respect to α, α', α'' we mean a group of mutually compatible individuals of $\mathfrak{M}$.

In order to partition $\mathfrak{M}$ into compatibility groups one may adopt the following procedure. (Its relation to the partition into groups without dissension discussed in the first note will be described later.) One associates to each member of $\mathfrak{M}$ any one of the codes that he accepts and then for each code, collects all of the members of $\mathfrak{M}$ with whom it has been associated. For some codes, that group may be empty because those particular codes are not accepted by anyone or because, though accepted by some, they are not associated with any of their adherents in the process of partitioning. For some other codes, that group may consist of a single individual, which is the case if the code, though possibly accepted by several persons, has been associated with only one of them in the procedure, so that this person constitutes by himself a compatibility group in the resulting partition. Clearly, the compatibility groups obtained by the prescribed procedure are mutually disjoint and jointly exhaust $\mathfrak{M}$. The selection for each individual of one of the codes accepted by him can be performed with various aims in mind; for example, to minimize or to maximize the total number of resulting compatibility groups; to equalize as far as possible the memberships of the various groups; to distribute the all-compatible individuals over the compatibility groups of the others roughly in proportion to the memberships of the latter.

The formalism of the procedure just described, which is at the root of many partitions, is simply this. To each element of a first class, there corresponds a nonempty part of a second class. (In the example just treated, the first class is $\mathfrak{M}$, the second is $\mathfrak{N}$, and the part of $\mathfrak{N}$ corresponding to an element of $\mathfrak{M}$ consists of the codes accepted by that member of $\mathfrak{M}$.) Then one somehow selects exactly one element from each of those nonempty parts and, for each element selected, collects all members of the first class to which it corresponds.

The constitution of groups without dissension can be subsumed under this formalism by choosing as $\mathfrak{N}$ a pair of attitudes denoted by + and −. With each member of $\mathfrak{M}$ there corresponds one of the three nonempty parts of $\mathfrak{N}$, that is, + alone, or − alone, or + and −, the last being symbolized by 0 in the first note. In case of two norms, the second class consists of four attitudes denoted by ++, +−, −+, −−; and to each member of $\mathfrak{M}$ there corresponds one of the following nine parts of $\mathfrak{N}$:

++; +−; −+; −−;
++ and +−; ++ and −+; −− and −+; −− and +−;
++ and +− and −+ and −−.

(In the first note, the last five are symbolized by +0, 0+, −0, 0−, and 00, respectively.)

Also the discrepancies between the behavior of a person and his wishes for the behavior of others can be investigated with respect to several norms. Here, we consider merely the case of two norms, N and N', of the type studied in the third note with the four categories:

$$(a), (b), (c), (d); \quad (a'), (b'), (c'), (d').$$

The two quadripartitions overlap yielding sixteen categories such as $(a) \cap (a')$, briefly (aa'), consisting of the persons belonging to both (a) and (a'); similarly, $(a) \cap (b')$ or (ab'). They are called *product categories* with the *factor categories* (a) and (a'), and (a) and (b'). We denote the sixteen product categories as follows.

$$\begin{array}{llll} (aa') = A, & (ab') = B, & (ac') = C, & (ad') = D, \\ (ba') = E, & (bb') = F, & (bc') = G, & (bd') = H, \\ (ca') = I, & (cb') = J, & (cc') = K, & (cd') = L, \\ (da') = M, & (db') = N, & (dc') = 0, & (dd') = P. \end{array}$$

The relationships between these sixteen categories are described in the following matrix, as were those between (a), (b), (c), (d) on p. 66. For example, the symbol u in the row of A and the column of E indicates that the members of A and E are compatible with respect to N and N'. The

	A	*B*	*C*	*D*	*E*	*F*	*G*	*H*	*I*	*J*	*K*	*L*	*M*	*N*	*O*	*P*
A	*u*	*u*			*u*	*u*										
B	*u*	*u*	*u*	*u*	*u*	*u*	*u*									
C		*u*				*u*										
D		*u*		*u*		*u*		*u*								
E	*u*	*u*			*u*	*u*			*u*	*u*			*u*	*u*		
F	*u*	*u*	*u*	*u*	*u*	*u*	*u*	*u*	*u*	*u*	*u*	*u*	*u*	*u*	*u*	*u*
G		*u*				*u*				*u*				*u*		
H		*u*		*u*		*u*		*u*		*u*		*u*		*u*		*u*
I					*u*	*u*										
J					*u*	*u*	*u*	*u*								
K						*u*										
L						*u*		*u*								
M					*u*	*u*							*u*	*u*		
N					*u*	*u*	*u*	*u*					*u*	*u*	*u*	*u*
O						*u*								*u*		
P						*u*		*u*						*u*		*u*

symbol u in the row of E and the column of E expresses that any two members of E are compatible. The absence of u in the row and column of C indicates that the members of C are self-incompatible.

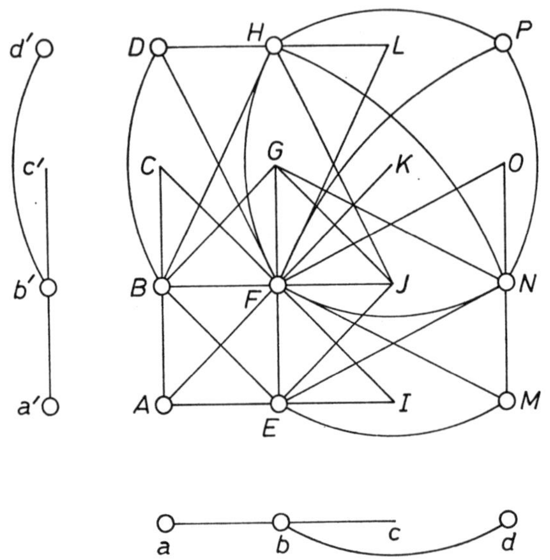

Fig. 5.

One readily verifies the following theorem, which holds not only for two but for any number, k, of factors. *In order that two product categories be compatible it is necessary and sufficient that each factor category of either one be compatible with each factor of the other.* In particular, *a product category is self-compatible if and only if all its factor categories are self-compatible and mutually compatible.*

The relationships between the product categories $A, B, \ldots, P$ can also be represented graphically (see Figure 5) using the factor categories as coordinates. The sixteen categories are represented by the points of a quadratic lattice (the nine self-compatible categories by small circles) having as coordinates the points (or circles) representing the factor categories. The representatives of any two compatible categories are joined by a segment of an arc.

The group of all self-compatible individuals (belonging to the nine categories A, B, D, E, F, H, M, N, P) can be partitioned into four compatibility groups:

A^+ consisting of A and any parts of B, E, F;
D^+ consisting of D and any parts of B, F, H;
M^+ consisting of M and any parts of E, F, N;
P^+ consisting of P and any parts of F, H, N.

If A^+ actually includes members of all four categories A, B, E, F, then no self-incompatible person can be adjoined to A^+ without A^+'s ceasing to be a compatibility group. There are, however, compatibility groups consisting of one member of C and any number of members of B and/or F or consisting of one individual belonging to G, one belonging to J, and any number of members of the all-compatible category F.

In the general case of k modes of behavior, α, α', α'',... $\alpha^{(k)}$, and the four categories of type (a), (b), (c), (d) associated with each, there are 4^k product categories, of which 3^k are self-compatible and capable of partition into 2^k compatibility groups.

The last matrices have a remote resemblance to matrices occurring in the theory of heredity. Of greater interest than this quite superficial similarity may be the question as to what can be concluded from the laws of heredity concerning the compatibility of the descendants of compatibility groups. (Of course it is clear that heredity is only one aspect even of those moral characteristics to which Mendel's laws can be applied and that environment and education must be considered.) We confine this study to the simplest case of a single norm and the two bipartitions of $\mathfrak{M}$ similar to those engendered by the norm of politeness. We thus assume that $\mathfrak{M}$ is divided (1) into the category p consisting of the adherents of N, and the category q of the others, and (2) the category r of those who demand fulfilment of N by their associates, and the category s of the others. We further assume that only four product categories are significant:

$$(a) = p \cap r, \quad (b) = p \cap s, \quad (c) = q \cap r, \quad (d) = q \cap s.$$

If Mendel's laws apply independently to the pairs p, q and r, s, and if intermediate characteristics can be disregarded, then obviously one of the following four cases must obtain.

	Dominant	*Recessive*
Case I	p and r	q and s
Case II	p and s	q and r
Case III	q and r	p and s
Case IV	q and s	p and r

If the members of the four categories are homozygous with respect to the characteristics p, q, r, s, then Table II shows the categories including the children:

TABLE II

Of a pair	(a,b)	(a,c)	(a,d)	(b,c)	(b,d)	(c,d)	(a,a)	(b,b)	(c,c)	(d,d)
In Case I	a	a	a	a	b	c	a	b	c	d
In Case II	b	a	b	b	b	d	a	b	c	d
In Case III	a	c	c	c	d	c	a	b	c	d
In Case IV	b	c	d	d	d	d	d	b	c	d

As has been shown in the fourth note, each partition of $\mathfrak{M}$ into compatibility groups consists of (1) some number, n_1, of groups of the *first type* each including members of (a) and possibly of (b); (2) some number, n_2, of groups of the *second type* each including members of (d) and possibly (b); (3) z groups of the *third type* each including one member of (c) and possibly members of (b), where z is the number of all members of category (c); (4) some number, n_3, of groups of the *fourth type* consisting of members of category (b). From the results summarized in Table II it follows that if only pairs of members of a compatibility group have children, then in each of the cases I, II, III, IV the groups of the first, second and fourth type are preserved. This means that each child of two members of a compatibility group of one of these three types (whether or not the parents belong to the same fundamental category) is compatible with both parents, all siblings, all members of the compatibility group and their children. For example, in each of the cases I–IV (thus under all circumstances), the children of two members of a group of the first type (whether they belong both to (a) or both to (b) or one to (a) and the other to (b)) belong either to (a) or to (b) and thus to the compatibility group of their parents.

But the children of members of a compatibility group are not necessarily homozygous; for example, the children of a member of (a) and a member of (*b*) in a group of the first type are heterozygous. Simple reflection shows that, even so, for any pair belonging to a compatibility group of first, second or fourth type not only the children but also the descendants in later generations are compatible with each other and with the other members of the group of their ancestral pairs. It is also clear that heterozygoty of the ancestral pairs does not affect this result if their recessive factors are compatible with the group or, as it may be expressed, provided that they are *recessively self-compatible*. From the point of view of heredity, *the groups of first, second and fourth type have preservation tendencies.* Environmental conditions probably will also contribute to the temporal constancy of these groups.

The conditions are altogether different in groups of the third type. A child of a member of category (*c*) and a member of (*b*) belongs to (*a*) in case I, to (*b*) in Case II, to (*c*) in Case III, and to (*d*) in Case IV. Only in Case II can such children be integrated into the compatibility group of their parents. Only in this case are there tendencies for the preservation of the group in the next generation. Tendencies to preserve the group in later generations, however, do not exist in Case II. For among the children of two heterozygous members of the second generation who both contain the factors *q* and *r* recessively, more than one may belong to (*c*). As to the other cases, apart from influences of environment and education, in Case I, the children of a self-incompatible person would leave the group; in Case III, they would be self-incompatible and might form new groups of the third type with members of category (*b*); in Case IV, they would not be tolerated in the group by the parent belonging to category (*c*).

CHAPTER V

LOGIC, IMAGINATION, REALITY, EVALUATIONS

A CONCLUDING LETTER ABOUT WHAT HAS BEEN SAID

The logico-mathematical notes on the compatibility of individuals that I have sent you develop the demystified ethics sketched in the epistemological notes. They carry the externalization of ethics to the point of, as it were, identifying moralities with the groups of those who adhere to them on the basis of their decisions. More generally, they associate to each value judgment a group of human beings, namely, the group of its adherents; or rather three groups of adherents of various types: the persons acting according to the judgment, those professing it, and those whose desires it reflects. You may call this externalized interpretation superficial; but it leads to positive conclusions, however modest. Exact thinking applied to ethics does not exhaust itself in mere negativism.

It is strange that so much of what fails to withstand rigorous criticism has been defended, especially in the so-called philosophy of values, while so little attention has been allotted to what can survive such critique. Only Kant gave primary consideration to the formal aspect of ethics; but on the one hand, he overestimated it and on the other, in the categorical imperative, he expressed it inadequately. His successors soon abandoned his thoughts, but they replaced them by writings which under the appearance of a greater positive content developed an obscure metaphysics that collapsed under epistemological criticism. Mere formalism was not sufficient for those men; this is understandable; but their own work did not fulfil even the demands of mere formalism.

Today [in the early 1930's], the logician contemplating logical or so-called logical thought in the social sciences, faces a strange spectacle. Some scholars definitely *overestimate* logic. Their hopes are too high; they look toward formalism for farther-reaching results than even a theoretical physicist would ever expect from logic and, in addition, for instructions in practical action. Scholars of a second type *misunderstand* logic. Wearing an austere mask they write lengthy dissertations on the relationships between normative and exact sciences, between jurisprudence and logic, and between what ought to be and what is – inserting a few terms that occur in exact discourses. But a closer scrutiny of their work reveals the lack of any sort of scientific productivity and, in many cases, even clarity. For example, observe the habit they have of joining logically unconnected statements by particles such as 'thus,' 'accordingly,' 'consequently.' I mention this petty detail, not out of pedantry, but because through such tiny holes even one with little logical training can glimpse the spirit behind the austere mask. Yet such writings, because of their ostensible precision, only too often bring exact thought into disrepute among nonlogicians. A third group of scholars, repelled by the pseudo-methodology of the second type, reject precise thinking in their field altogether. They are under the misapprehension that it destroys all that is positive – an opinion that they often try to buttress with arguments which, were they justified, would condemn even theoretical physics. These men certainly *underestimate* logical thought and by neglecting what epistemological criticism does sustain are themselves guilty of damaging something positive. They also overlook the fact that contempt for the formal can be taken seriously only in one who has mastered formalism. Altogether, this leaves relatively few scholars in the area of the social sciences who bring to exact thought that balanced judgment which arises from a command of logic.

Should you consider the course set in these notes merely as a logical game or as "scholastic sophistries" which you don't care to regard as ethics, then I repeat: you are free to classify or designate these theories in any other way; for example, by a sarcastic name, such as "ethistics," just as the logic developed by the mathematicians is referred to, especially in German, as "logistics" – a term which to some philosophers has a derogatory ring. As for myself, I would be flattered if the relation between an exact theory of human associations and traditional ethics were to be compared with that between logistics and traditional logic. For you certainly know my opinion of the logic that mathematicians have developed since the middle of the 19th century and of the nonmathematical logic professed during that period. And as far as the scholastics are concerned, while I am of course aware of their weaknesses it is precisely some of their so-called sophistry that makes me think that few groups of scholars have been treated as unjustly as the great medieval logicians.

To be sure, the statements about groups and relations in my notes are without exception *simple*. But simplicity in my eyes is never a fault; and in this particular case, some statements are so utterly simple that they only have to be proposed in order not again to disappear from ethics.

It might also be said that in searching for the logically positive elements of ethics one is actually led to theories of human behavior that transcend what is specifically ethical in the traditional sense. For groups similar to those considered in my notes might also be formed according to extra-ethical, especially aesthetic, political or economic criteria (groups of the last kind, incidentally, might be not irrelevant in theories of economic action; nor is it an accident that in the famous dialog that begins with an examination of ethical concepts Plato soon allows Socrates to take up problems of the State). But can a method be faulted for being able to accomplish more than what it was destined to achieve? All I would concede is that ethistics be included in a theory designated as "sociologistics."

Though not denying all hope for the course set by these logico-mathematical notes, you may feel that such ideas can lead only to very *little*. But just a few principles have yet been stated, and these matters are greatly expansible; thus it seems to me that an altogether pessimistic view is premature. At any rate, apart from the establishment of facts and of empirical regularities, I can see no way other than the development of a logic of ethics by which ethics can achieve *anything at all*. So my notes about compatibility groups may be a beginning. For in the literature, the only studies of groups that I have discovered proceed in directions different from mine; and even such studies I found in very few writings; the fine works of G. Ratzenhofer and L. von Wiese deserve mention.

In addition to being the logical development of the ideas presented in the epistemological notes, the theory of compatibility groups enables one to make assessments of various traditional ethical theories. As an example, consider Nietzsche's ideas. As mentioned in one of the first notes, the coexistence of master and slave laws can lead to concrete precepts for the individual only if supplemented by a precise criterion indicating who is master and who is slave – a point on which Nietzsche never clearly expressed himself. A practical criterion would be the judgment people make when setting up the double code: according to whether a person has declared himself a master or a slave at that time he would always be regarded as such. But what about the compatibility groups during the period of the creation of the double code? If it is proposed that individuals with particular external characteristics (for example, of parentage, intelligence, physique, achievement) should be granted a favored position, then certainly also some persons who lack those characteristics and have no claim to favor would agree to the proposition, especially if there were a chance for them or their progeny to acquire the characteristics and thereby eventually the status of masters. Even more people would agree if they were to realize that they might somehow indirectly benefit from the favors shown to the individuals with the characteristics in question: if, for example, the favored position would go to persons who had made generally useful inventions or discoveries and from whom further results could be expected with greater likelihood if they were granted special privileges. If, however, it is proposed that the favored status go to the persons who simply characterize themselves as masters, then these individuals might find a few admirers to assume the role of their slaves, but most people would doubtless take one of the following two positions: either without sensing any special personal qualities of superiority they would designate themselves as masters in order to participate in the benefits of that position; or, mustering the courage to be honest, they would state that they feel neither destined to be masters nor willing to be slaves of persons who simply declare themselves masters. If the former attitude were to prevail, then the number of masters would be so large that the exploitation explicitly favored by Nietzsche would fail for lack of victims. If the majority were to take the latter position, then the Nietzsche State would be relatively small while the majority of people would arrange their lives in different ways. But the objectives of the masters would be achieved neither in the case of too many masters nor in a double-standard state that is too small; in the latter case, the masters might well be worse off than are the normal members of other communities.

I wish to conclude these remarks about logic with a few words about the relation between the logic of ethics and what you have called the psychology and biology of morality. Clearly, the latter supply material for the former but cannot invalidate conclusions of the logic of ethics; just as the logic of ethics, in turn, cannot create or refute facts of the psychology or biology of morality. Conceivably, these sciences may bring to our attention new forms of organizing man's ethical and legal life – forms of organization as yet unimagined, which could also direct the logic of ethics toward new subjects and relationships. As things stand today, however, it is the logic of ethics that seems more suited and more likely to be a point of departure for basically new ideas and suggestions.

Let me now for a moment set logic entirely aside and use pure *imagination* in a thought-experiment founded on a synthesis of my notes (the notes being, of course, independent of what follows).

Imagine a large class of people gathering to discuss basic problems of their coexistence. Various regulations are proposed and compatibility groups emerge. Each group adopts rules to which each member of the group assents in all points important to him. These groups will be referred to as *voluntary associations*. You will remember that they are not necessarily groups of consentience or even groups without dissension. Consentience and even lack of dissension in judging moral questions are not prerequisities for mutual compatibility. Compatibility groups may include mutually nonconsentient as well as self-incompatible individuals. Thus compatibility groups are in general more extensive than groups of consentience, and the incorporation of disjunctive norms in codes might make them even larger.

But the population changes through birth and death; and some individuals alter their decisions. For compatibility groups and voluntary associations to be permanent as such, their codes must take into account potential changes in the decisions of members, such as those typical of advancing age; and these codes must furthermore allow for amendments or for the creation of subgroups to recognize parallel changes in the decisions of large numbers of individuals. Inter-group agreements (in harmony with the individual codes) must regulate the classification of those entering adulthood and the reclassification of some adults. On the other hand, there may appear hereditary as well as environmental tendencies toward the preservation of groups.

Some codes entail limitations concerning the number of individuals who may espouse them; in particular, some regulations require a minimum number of adherents in order to become viable at all. Clearly, the person who desires to join such a group must make more concessions about matters of only secondary importance to him than must an individual who intends to join a smaller group. Equally clearly, if the advocates of a plan agree about major issues, an argument over details could lead to a rift endangering the practicability or at least the effectiveness of the entire plan. There might remain to such a group only the choice, familiar from party politics, between renouncing less important claims for the sake of unity and jeopardizing important ones by seceding.

The question as to whether unlike codes can coexist in one common territory must be decided in each particular case on its own merits. But it does not seem to make undue demands on the imagination to conceive indeed of the coexistence of diverse codes for diverse groups of inhabitants of the same territory. In many cases such arrangements would only presuppose that the members of the various compatibility groups be recognizable as such, while technology would facilitate secessions in some extreme cases.

"But what good are fantasies that result in constructions foreign to *reality*?" you will probably ask. "What is the use of basing a thought-experiment on attitudes toward general matters about which the overwhelming majority of people actually take no position at all, except perhaps through the initiative of prominent individuals?"

Well, leaders who make decisions that they wish to realize and who know how to influence people are always present; ideas for feasible systems of norms to regulate human behavior certainly exist; and each of these ideas, even if only through the initiative of a leader, is an object or at least a possible object of the will of masses. But these ideas as well as decisions of various leaders not only *deviate* from each other – some of them to the point of utter incompatibility – many decisions, even some of those treated in traditional ethics, *refer* to one another.

Let me say, first of all, that the existence of those mutual references is one of the reasons why in talking to you I have emphasized that the analogy between ethics and geometry is not very far-reaching. For axiom systems of the various geometries coexist without any mutual references, so that even total diversity of the propositions of various geometries does not create any conflicts. Contradictions would only arise if one were to mix the statements of several systems or to line them up into a single more comprehensive code, and this, of course, it never occurs to anyone to do.

What is the actual regulation of human behavior and social relationships in the face of interacting resolutions that are based on conflicting decisions? Humanity is divided into groups, often referred to as societies, each of which consists essentially of the inhabitants of some region and has its laws and regulations. Violations of laws are penalized in order to create motives against transgressions. Except for occasional interference of inter-societal relationships, the preservation or modification of laws and regulations is a question of the distribution of power within the society. Changes come about in one of two ways: either some individuals *force* their decisions on all – 'some' being used here in the broadest sense to include single persons such as a founder of a religion, a reformer or a dictator as well as a powerful minority or even the majority of the people; or there is a *compromise* in the sense that each of several groups gains the acceptance of some of its decisions. The outcome of such struggles is influenced to a great extent by coincidences – elements that have nothing to do with the content of the laws or with people's positions regarding them.

How does this reality relate to the thought-experiment that I have described? Would the constitution of compatibility groups and voluntary associations be a means to obviate the collisions between human decisions? It is clear that the answer is negative in many cases, in fact, for whole classes of behavior. It will be one of the most important tasks of a detailed logic of morality to classify the modes of behavior accordingly and to examine the characteristics of the decisions that can, and of those that cannot, be integrated into a system of compatibility groups.

The main stumbling blocks are to be found among decisions which refer to one another. If it is the intention of the members of one group to subjugate the members of another group against their will or to take possession of the territory inhabited by the latter – intentions of which political history offers examples in abundance – then it is clear that the conflict cannot be eliminated by the constitution of voluntary associations. It can be ended only in one of two ways: by a change of at least one of the conflicting decisions or by struggle and violence, the results largely depending on coincidences.

The clash of decisions prompting the War between the States provides an example in which idealistic, political and economic elements interacted. The whites in the southern States were in favor of slavery, the whites in the northern States were against it. As far as southern whites were concerned this might have been a conflict between two clearly defined and even spatially separated decisions which could have been resolved by a separation of the two groups as demanded by the southern States. Of course, taking into consideration also the blacks in the South, most of whom naturally were against slavery, it could be predicted that after a secession the southern States would not form a compatibility group. Moreover, the northern States were not only opposed to slavery in their own territory, where it did not exist, anyway; but for idealistic and economic reasons they demanded the abolition of slavery in all the United States together with the preservation of the Union. Each group was thus faced with the dilemma of abandoning its decision or pursuing it by force.

A similar example is presented by the collisions which would arise from what Nietzsche called the "will to power." To a large extent the underlying decisions belong to those that concern the very partitions into groups. Also if one nation has adopted as a norm to force its own religion on another nation a decision of a higher logical type has been made – a decision of one group which concerns the decisions of another.

I do not believe, however, that the situations described in these examples are present in *every* case where legal or moral precepts are in question. If a person wants to be sure to find a certain type of behavior in those around him, it is sufficient for him to try to establish precepts guaranteeing the desired regularity in his environment. It is not at all necessary for him that the precepts become laws for everyone – for the simple reason that he will never have anything to do with the vast majority of men; and even within his community all he really wants to achieve is that a particular type of behavior be observed where he himself is concerned. Yet the traditional method of establishing moral and legal precepts requires the institution of general regulations. But the traditional mechanism of legislation is abandoned in the thought-experiment. It is replaced by the restriction of the validity of a norm to the group which decides to accept it.

In many cases, a group of people who wish to observe a particular mode of behavior need, for the realization of this wish, certain legal guarantees within their group: yet it is irrelevant to them if groups with other views behave differently since even in coexisting groups the different modes of behavior do not exclude one another. Under the traditional conditions, however, the first group must try to enforce universal laws that conflict with the wishes of the second group, whereas the latter would try to abolish such general laws even though they may have no interest in forcing their own will on the first group.

Of particular importance in this connection is the fact that certain plans cannot be executed without *some* regulation and binding norms for the adherents. There are consistent ideas, among them some of momentous importance, that cannot be carried out either without any regulation or under universally valid norms – plans that can only materialize through the actual constitution of voluntary associations.

There remains the question of the practicability of voluntary associations. If at first you should be pessimistic, then consider that in past centuries (and actually in a not too distant past) people found the coexistence of various religions completely unthinkable, while today groups with unlike norms in this respect are a reality. There are other examples pointing in this direction such as special class moralities to which members voluntarily subject themselves and religious orders which are freely joined. The domain of modes of behavior thus regulated can undoubtedly be expanded.

To be sure, the question of *evaluation* that you have broached has not bccn answered in my notes. Up to this point, also the present letter has been entirely devoted to matters about which, in accordance with your initial warning, I have strictly avoided subjective statements. And even here in conclusion, I do not propose to formulate any evaluations in the ordinary sense: I will not take a position for or against any particular moral idea. Only about the organization of various ideas and systems of ideas (thus about what might be called *morality of a second level*) will I venture to express what amounts to a value judgment: the wish that, wherever possible, the conceptual scheme of compatibility groups be realized in voluntary associations.

There is no proof for the superiority of this scheme. Just as on the first level of morality, so too on the second level I consider any demonstration of necessity, naturalness, justice or superiority to be illusory.

Yet you will probably ask *why* I have expressed that particular desire – the wish for the constitution of voluntary associations. I therefore will summarize those special features of that desired form of organization which correspond to certain of my traits so as to awaken just that wish.

A first characteristic of the plan is the open recognition that the ideas governing each group are determined by its voluntary decisions. Even though this is a rather superficial feature it affects my overall judgment by a two-fold appeal: I prefer decisions to be frankly proclaimed rather than to be hidden; and I have a strong aversion for empty phrases. Now in my opinion the cloaks that are often wrapped around moral decisions are indeed empty phrases – verbiage that in matters of such vital importance is not only distasteful to me but dangerous. For phrases that say nothing tend to obscure what *can* be said; and this includes: informative statements that may render the consequences of various moral ideas more apparent and the choice between them more conscious; the establishment of facts and regularities; even occasional logical considerations. It is for these reasons that a clear and open emphasis on decisions is attractive to me.

Another, more important, characteristic is this. Wherever the conceptual scheme is practicable the decision of each group can be carried out with the full energy of that group and a minimum of inner friction. In contrast, if the will of a mere portion of a group or a compromise between the wishes of several groups or subgroups becomes the rule for all, then these precepts are not acknowledged by some or not fully acknowledged by anyone. They consequently will probably be resisted or circumvented by some individuals or at best be carried out without general enthusiasm; and the plan on which the regulation is founded will never materialize with the support of all available forces. Restricting the execution of an idea to the group of its adherents gains in intensity what it loses in extension.

There is a third characteristic. Here I ask you to consider my words closely; for notwithstanding the importance of the matter, I will state it only briefly. If various groups work under a common system at cross-purposes, then even some of the most important features of the underlying ideas may fail to materialize so that their effects may never be clearly seen or the basic plan fully judged. A system of several self-imposed regulations of various groups makes the effects of their diverse ideas manifest. Only on this basis have moral plans, removed from primary passions and from the happinstances of many struggles, the opportunity to prove themselves. Only in this way do ideas and their effects become comparable with one another. Only the organization proposed in the thought-experiment opens the possibility of a contest of plans.

The categorical imperative is inadequate for the universal regulation of our actions. But in connection with his explanation of morality, Kant wrote also another sentence that is remarkable, though less famous than his imperative; and this sentence becomes a reality only in voluntary associations: "Laws ought to be found inherent in every reasonable person and must be able to arise from his will."

Whether you now will say that of your initial warnings I have observed the first by not including any evaluations among my ethical discussions, but have ignored the second by treating morality in a purely intellectualistic way – that I must leave for you to decide. For my part, I believe that I deserve this reproach less than any other philosopher. For one of my main objectives in these notes has been precisely to relegate intellect and reason to their proper, secondary role in ethics; and to replace pretended results of pseudo-logical thought, which cannot withstand epistemological critique, by a voluntaristic foundation of morality upon decisions. But be that as it may, let me emphasize once more that neither our conversation nor my notes have advanced a single personal value judgment; and even here in conclusion I would not have touched on such matters, had you not asked about them. What I really felt compelled to do was to share with you and others a few results of exact thinking about the facts of morality and organized coexistence.

Vienna, Winter of 1933/34

POSTSCRIPT TO THE ENGLISH EDITION

I

This book was written in Vienna in the winter of 1933/34 – during the period before the development of linguistic philosophy, of the existentialism and the behaviorism of the 1940's, of mathematical models in sociology and of experiments in the social sciences. Its philosophical background can be described in a few paragraphs.

Traditionally opposed to Hegel, the Viennese philosophers studied Hume and Kant, and read Nietzsche and Tolstoy. Earlier in the century, the Austrian philosophical scene had been dominated by two schools: Franz Brentano's and Ernst Mach's. The 1920's witnessed the rise of the group called in Vienna the *Schlick-Kreis* after its founder and known in the world as the *Vienna Circle*. All three groups were primarily interested in epistemology rather than in ethics.

Brentano's ethical ideas are epitomized in sentences such as "Good is that which is worthy of love or what ought to be loved," "Evil is what it is impossible to love correctly," "What is worthy of love one recognizes by intuition." While Brentano's disciples hailed these phrases as the results of deep insight I have never discovered their cognitive significance. Nor have I been able to derive information from Max Scheler's and Nicolai Hartmann's characterization of values as essences given by immediate intuition.

Mach's positivism and his antimetaphysical attitude profoundly influenced many scientists but had no comparable impact on writers of ethics. Fritz Mauthner, the critic of language, was one philosopher who as a student in Prague was deeply impressed by Mach. He wrote at the beginning of this century, "Morality is only in words," and criticized ethics for "building its assertions in terms of unclear words of the common language and ending up with monstrous phrases such as 'I ought to will'."

The members of the Vienna Circle had originally intended to develop a synthesis of Mach's positivism and the logic-oriented philosophy of

Bertrand Russell, whence they characterized themselves as logical positivists. In the early 1920's, the group became acquainted with Ludwig Wittgenstein's *Tractatus Logico-Philosophicus*. When I returned to Vienna in 1927 after an absence of more than two years and was invited to attend the meetings of the Circle I found that the logical and epistemological parts of the *Tractatus* had deeply impressed M. Schlick, H. Hahn, R. Carnap, F. Waismann and H. Feigl. Not so the short concluding part of the book with sayings about that which, according to Wittgenstein is unsayable – the part that includes brief references to ethics. The restriction of science to value-free statements had always been part of the program of the Circle, so that the members found the sentences, "In the world everything is as it is and happens as it happens. In the world there is no value" congenial, if not especially illuminating. On the other hand, they considered as eccentricities statements such as "The meaning (*der Sinn*) of the world must lie outside of the world," "Not *how* the world is is the mystical, but *that* it is," "There is indeed the ineffable. It evinces itself; it is the mystical." These aphorisms were ignored – except by O. Neurath in whom they created a deep distrust of Wittgenstein as a mystic.

In the Circle, I don't remember having ever attended discussions of ethical problems. In my recollection, Schlick was the only member who at that time pursued studies in ethics. Before I came to Vienna he had written a beautiful, poetic booklet on the meaning of life[1] in which he extolled youthfulness and play. In 1930, he combined these thoughts with a critique of Kant's ideas on absolute norms and duty in a book, *Problems of Ethics*[2] – a more systematic work, which yet exhibited only a modicum of logical analysis. Though written at a time when Schlick's epistemological views were strongly influenced by Wittgenstein that book on ethics mentions neither him nor the *Tractatus*; and from Waismann's posthumously published notes[3] it also appears that ethical questions played only a minor role in his and Schlick's discussions with Wittgenstein over the years.

[1] M. Schlick, *Vom Sinn des Lebens*, Berlin, 1927.
[2] M. Schlick, *Fragen der Ethik*, Vienna, 1930; *Problems of Ethics*, New York, 1939.
[3] F. Waismann, *Wittgenstein und der Wiener Kreis*. Aus dem Nachlass herausgegeben von B. F. McGuinness, Oxford, 1967.

II

Should there be no room at all, I asked myself, for exact thinking in the field of ethics?

A first survey looked unpromising. There are but few and rather trivial logical relations between precepts; for example, the norm "Thou shalt not steal" may be said to imply norms such as "Thou shalt not steal a horse" and "Thou shalt not steal a horse or a car." But such trivia are not what is in the minds of writers who claim that they can deduce rules of conduct from the precept that everyone should follow whatever happens to be the writer's ideal – the interest of a community or the good of mankind or harmony with nature or justice. An analysis of what had been called deductions, however, forced me to conclude that those authors were not sufficiently familiar with logical processes. When I looked for specific rules of conduct that general ideals might imply in a strict logical sense – and what logical sense is there but a strict one? – I could not find any. Correct deductions of more specific rules (for example, of commandments of the biblical decalog) from the precept that one should follow one of those traditional ideals are impossible.

I then tried to describe the common use of the words 'good' and 'just' tracing them to the way by which people learn to use them in childhood, and emphasizing the role of rewards and punishments in the formation of that usage as do, of course in a more refined way, behaviorists today.

Greater importance I attributed to an examination of Kant's categorical imperative. In his book, Schlick had criticized Kant's "absolute ought" as "a demand without a demander," comparable to "an absolute uncle without nephews and nieces." I proceeded pragmatically, just as I had in the study of the ideals, by asking myself what specific precepts follow from the categorical imperative in a concrete situation; in other words, exactly what it is that Kant's demand demands. I found that in most cases several mutually incompatible types of behavior are compatible with Kant's principle. Thus, like the ideals, the categorical imperative fails to provide a basis for the regulation of a person's conduct; and this even under the assumption that everyone intends to comply with the rule – not to speak of cases where there are intentional violations of the imperative by other persons.

But if there are no cogent premisses for rules of behavior, what *is* the basis of a person's choice of a specific mode of conduct or of a particular ethical code?

The answer given in this book in 1934 is that a person's conduct in a specific instance and his or her adoption of some moral code have, in the last analysis, no basis other than the person's *decisions* (*Entschlüsse*). This is also true for actions performed in obedience to an authority that penalizes transgressions, and for behavior modelled according to a religious code which derives its norms from a tradition of its superhuman origin. (In the latter case, the decision lies in the adoption of the particular faith.) It is ultimately true even for a type of conduct that is of paramount importance among individuals educated in a civilized society – commissions and omissions said to be dictated by the voice of conscience.

Emphasis on decisions also plays a major role in the post-World War II existentialism. But existentialists base their ethics on philosophical discussions of existence and on a metaphysics of "being and nothing," while the results of this book are (and make a point of being) totally free of metaphysics. Nor will the reader find any implicit remnant of Kant's categorical imperative as may be present in the existentialist thesis that by deciding his or her conduct a person makes decisions for *everyone*. All that can here be admitted is the obvious remark that a person's decisions or, rather, his ensuing actions and words are *examples* for everyone or, more precisely, for everyone who learns about them; and beyond this, they do not prejudice the decisions of others.

Must exact thinking in the field of ethics really halt at these negative results and retreat into an analysis of the various uses of the words 'good' and 'evil' if it is not to be engulfed by mysticism? Must it confine itself to the epistemological critique of value judgments, which essentially amounts to repeating the countless times proclaimed thesis, "Nothing that *is* implies anything that *ought* to be"? Must it, beyond this, resign itself to the statement that value judgments ultimately rest on decisions? And even *this* statement has little positive content; while implicitly admitting that actions are garbed in verbiage the statement ultimately amounts to saying little more than that people act the way they act. (Although these epistemological analyses of ethical material lack positive content their potential importance must of course not be underestimated. If these critiques were to be taken seriously, especially by all young people, then private counselling, politics on all levels, and international relations would in the future be free from the empty verbiage and the hypocrisy that have plagued them in the past.)

Moreover, there are numerous and diversified *empirical* studies of great importance connected with moral decisions: anthropological and historical descriptions of habits and precepts in various parts of the earth and of their development in various epochs; biological and psychological investigations; sociological and economic results that may be advanced in support of some decisions and as arguments against others. All this, however, did not touch the basic question that was occupying my mind in 1934: Is there a possibility of applying *logico-mathematical* thinking to ethical material?

Certainly, I have always felt, the answer is negative in those cases where ethics is most frequently invoked – those many conflicts of conscience that are important in human lives and highlighted in novels and dramas. Ought a person in a dilemma to take action α or action β? Often the precepts to take these actions are believed to be consequences of more general rules, A and B, respectively. But be that as it may, all a moral philosopher can do is to express *his* decisions concerning α and β or A and B, and to tell what in a similar dilemma he himself did or would do or, at any rate, wishes others to believe that he would do. What he says may greatly comfort persons in the dilemma or strongly influence them. With cognitive processes, however, these and all other moral judgments of investigators have nothing to do.

But could exact thinking be utilized, if not in dealing with personal dilemmas and conflicts of conscience, then in the treatment of social problems and conflicts of masses? Of course in this direction too, if results were to be achieved, evaluations – even strongly felt value judgments – should be strictly withheld from their presentation.

III

In fervent attempts to demonstrate that exact thinking is not entirely doomed to ethical negativism it appeared to me that a basis for a positive approach might be obtained by what in this book is called the *externalization of ethics*. To any norm, any moral code and, more generally, to any value judgment I associated the group of its adherents – more precisely, three (by no means necessarily identical) groups: of those who shape their actions according to the judgment; of those who profess it orally; and of those whose wishes are in accord with it. The members of the first two of these groups are defined by overt characteristics and are directly observable as such; confidential questionnaires may in some cases reveal at least the size of the third of these groups.

This externalization of value judgments is reminiscent of the representation of the father-son relation among humans as the class of all pairs of men the first of whom is the father of the second. From this "extensional" point of view, many modern logicians represent each binary relation as the class of all pairs so related or even *define* the relation to be that class. Still closer to the externalization of value judgments is the identification by some logicians of a property with the class of all elements sharing that property.

That such an externalization might appear to some philosophers to be a very superficial treatment of ideas was clear to me. But the group of adherents certainly is thc most prominent observable correlate of a code; in fact, besides written and oral formulations it is its *only* observable correlate. And the introduction of classes of persons corresponding to value judgments and norms indeed opens a field for exact thinking.

A first remark concerns a certain reciprocity or duality between individuals and norms. There is an analogy to plane geometry, which studies the class of all points on a line and the class of all lines on a point. But whereas a point and a line are either incident (i.e., on one another) or non-incident, the relation between a person and a norm lends itself to a more detailed description. Besides approval and disapproval, there is indifference, and even further specifications would be possible: one might also consider uncertainty of an individual as to whether he is approving or indifferent and whether he is disapproving or indifferent; and, more generally, various degrees of importance that an individual attaches to a norm. Moreover, all these distinctions can be made on each of the levels of actions, words, and wishes. A person's approval or indifference or disapproval is manifested in actions by his always or only sometimes or never fulfilling the requirements of a norm.

Disregarding these refinements one can say that (1) each individual determines a tripartition of the class of norms: into those of which he approves; those to which he is indifferent; and those of which he disapproves; and (2) each norm determines a tripartition of the class of all individuals into those approving of the norm; those indifferent to it; and those disapproving of it. And further reciprocal concepts can be introduced on both sides of the ledger. Reciprocities are, however, of only secondary importance.

The main applications of exact thinking are to *relations between individuals and groups as well as between groups and groups – groups and relations based on people's decisions and their demands of others.* An example that is closely related to cases treated in this book may illustrate these studies. It is taken from a paper[4] that I published in 1938.

Consider a class of people which is divided into two disjoint groups, G_1 and G_2. The groups may consist of the altruistic and the selfish individuals, respectively, or of the members of two nationalities or of two religious faiths. Assume that each member of the class has exactly one of four attitudes about his associations with others: he wants to associate either only with members of G_1 or only with members of G_2 or with everyone or with no-one. Members of the last two groups will be said to be *tolerant* and *unsociable*, respectively. The four attitudes divide the group G_1 into four disjoint subgroups, G_1^1, G_1^2, G_1^{12}, G_1^0, respectively. Analogously, G_2 is divided into four subgroups G_2^1, G_2^2, G_2^{12}, G_2^0. These eight will be referred to as the *fundamental* groups. In many situations, some of them are empty.

In each of the four groups G_1^1, G_1^{12}, G_2^2, G_2^{12}, any two members are willing to associate with one another. Groups of this kind will be said to be *cohesive*. The other four fundamental groups are non-cohesive; in fact, for each of them any two of its members are unwilling to associate with one another; their members may be called *self-incompatible*.

What cohesive (not necessarily fundamental) groups can be constituted? In the following enumeration, for any two groups, A and B, the symbol $A \cup B$ denotes their union, that is, the group of all individuals who are members of A or of B or of both. Each cohesive group belongs to one of

[4] 'An Exact Theory of Social Groups and Relations', *Amer. Journal of Sociology* **43** (1938), 790. Cf. also my lecture 'Einige neuere Fortschritte in der exakten Behandlung sozialwissenschaftlicher Probleme' in the booklet by E. Späth, H. Thirring, H. Mark, W. Heisenberg, and K. Menger, *Neuere Fortschritte in den exakten Wissenschaften. Fünf Wiener Vorträge*, Vienna, 1936.

the following five possible types (three of which have two subtypes):

(a) *pure* groups: (a_1) subgroups of $G_1^1 \cup G_1^{12}$ or (a_2) subgroups of $G_2^2 \cup G_2^{12}$ (among them are the subgroups of any one of the four cohesive fundamental groups);

(b) *tolerant* groups: subgroups of $G_1^{12} \cup G_2^{12}$ (subgroups of either G_1^{12} or G_2^{12}, being both tolerant and pure, are the only cohesive groups belonging to more than one of the five types);

(c) *centered* groups: (c_1) one member of G_1^2 and a subgroup of G_2^{12} or (c_2) one member of G_2^1 and a subgroup of G_1^{12};

(d) *singular pairs*, consisting of one member of G_1^2 and one of G_2^1;

(e) *hermit* groups, each consisting of (e_1) one member of G_1^0 or (e_2) one member of G_2^0.

The situation can also be represented graphically. Figure 6 shows (the symbols of) the fundamental groups – the cohesive groups encircled. A line joins two groups if and only if any member of either group is willing to associate with any member of the other.

Each person by himself constitutes a possible group, called a *singleton*, which is (in a trivial way) cohesive. Besides hermit groups, there may exist singletons of types (a) and (c). Each of the former consists of one member of any of the four cohesive fundamental groups; each of the latter, of a self-incompatible person (and an empty subgroup of tolerant people). The hermit groups are singletons that are inextensible (i.e., not parts of any larger cohesive group).

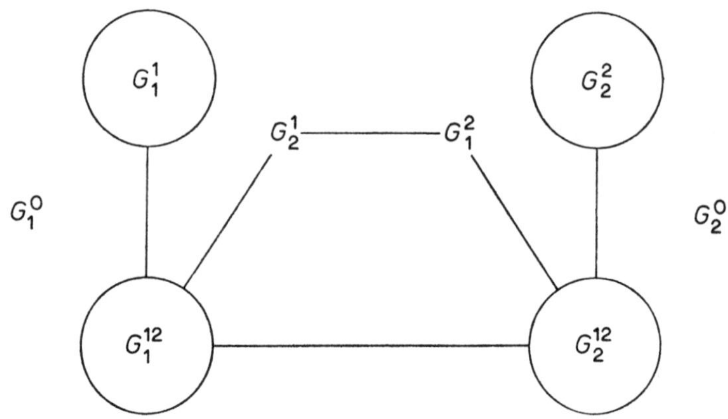

Fig. 6.

If G_1 and G_2 consist of the members of two faiths, then the groups G_1^2 and G_2^1 are empty. If the members are of two nationalities or of different ethnical extraction, then there may also exist snobs belonging to G_1^2 and/or to G_2^1. If G_1 and G_2 are the men and the women at a party, then, with respect to the association of dancing, all four cohesive fundamental groups are empty. The only cohesive groups are singular pairs (dancing couples), hermits (non-dancers), and other singletons (wall flowers of either sex).

If G_1 and G_2 consist of the altruists and the selfish, respectively, or of the slaves and supermen in Nietzsche's sense, then in either case G_2^2 and G_2^1 are empty; and only four types of cohesive groups are possible: groups of type (b), in which the supermen exploit the slaves; groups of type (c_1) consisting of slaves and one dictator; hermit groups; and groups of type (a_1) (which Nietzsche seems to have overlooked or certainly underestimated) consisting of altruistic people without parasites.

Some prehistoric societies are reported to have had the following structure, G_1 being the group of women and children, G_2 the group of adult males: the fundamental groups G_1^2, G_1^0, G_2^2, G_2^{12} were empty; and only three types of cohesive groups existed: groups of type (c_2), that is, families with one *pater familias*; groups of type (a_1) consisting of members of G_1^{12} who flee the tyranny of such families; and groups of type (e_2), each consisting of a single male who aims to become the center of a family.

Kant's categorical imperative demands that a person follow a norm if he can wish that it be followed by everyone. If N_1 and N_2 are two different norms regulating the same situation, and if G_1 and G_2 are the groups of their respective adherents, then it is the members of the groups G_1^1 and G_2^2 who follow Kant's precept. But members of G_1^1 are incompatible with members of G_2^2, whence the individuals following the imperative cannot form one cohesive group. On the other hand, centered groups and singular pairs are cohesive even though their members do not follow Kant's precept. Hence *obeying the categorical imperative is neither a sufficient nor a necessary condition for compatibility and the constitution of cohesive groups*. Thus apart from the failure to imply unique material precepts (mentioned in Section II), Kant's principle faces difficulties even from a purely formal point of view.

In the decades since this book first appeared it has become customary to speak about *models* of social phenomena: the elements of a class are assumed to be interrelated in a way that reflects some aspect of a social phenomenon or a simplified version thereof. The preceding discussion has dealt with a model in this sense, namely, a model of certain human relationships. The assumptions are that a class of individuals is divided (1) into two disjoint groups on the basis of possession or nonpossession of a certain (unspecified) characteristic, and (2) into four disjoint groups in which the interrelations reflect the willingness or unwillingness of individuals to associate with others on the mere basis of their possessing or not possessing the said characteristic. (This, incidentally, is the point where the preceding model greatly simplifies the social reality. There, most forms of voluntary association are based on the individual's possession or nonpossession of *several* characteristics. But one must beware of attributing the afore-mentioned difficulties of the categorical imperative to this simplification. Without it, those difficulties would in fact be aggravated.) One of the simplest theorems concerning the preceding model (that is, one of the immediate consequences of the assumptions) is the enumeration and description of all possible types of cohesive groups.

In current parlance, parts of the book apply mathematical reasoning of utmost simplicity to sociological models. In 1934, it should be remembered, many if not most of the sociologists and philosophers in Central Europe regarded some of what today is even elementary school mathematics as esoteric mysteries.

What a theory of sociological models could ultimately achieve was, and still is, hard to foresee. But in any case, the situation as it appeared to me when writing the book was this. The many historical and ethnographical, biological and psychological, economic and sociological facts that have been alluded to may of course *influence* decisions in various degrees; but they do not concern the decisions proper. In an investigation of the latter, if one completely eliminates value judgments – implicit as well as explicit, hidden as well as open evaluations! – *a study of the relations between the groups of their adherents seems to be all that can be done.*

IV

Now consider an actual population in which several groups adhere to mutually incompatible decisions.

One of the traditional ways of coping with this situation has been by *dictate* in the sense that a norm or a code is forced on nonadherents. This method has a wide spectrum of forms: the dictate may be imposed only on some of the nonadherents, or extended to all of them thereby resulting in a monistic regulation; it may issue from a single individual, from a small group, from a sizable minority or from the majority. Regulations by majority dictates have been practiced, under the name of democracy, since the flowering of ancient Athens. By contrast, under tyranny and, in this century, fascism, the dictate issues from one person or a small group.

The various forms of dictates, for all the practical importance of the variances between them, from a logical point of view differ only in degree. A qualitatively different method of handling the situation is by *compromise*: either by a modification of norms that leads to mergers of the groups of their adherents or by combining (unmodified) norms of various codes into an eclectic system. (There are even eclectic religions.)

Both dictates and compromises aim at and result in *uniform* regulations for the entire population – but at a high price. Dictates satisfy only some individuals; compromises satisfy either no-one fully in any controversial matter or everyone in some respects while not in others.

Evaluation-free studies cannot result in norms concerning the organization of groups any more than they can characterize good or just or right decisions of individuals. All they can do in a practical way on *any* level is to point to possibilities some of which may have been overlooked or neglected. It is in this sense that the ideas set forth in this book suggest an alternative to the uniform regulations discussed above. The study of human groups and interrelations that are based on decisions of individuals and their demands on others points to the formation of voluntary associations, the constitution of mutually compatible cohesive groups in peaceful coexistence, and the restriction of norms and codes, as far as possible, to the body of their adherents – briefly, the study suggests *pluralistic* solutions. Though the *term* 'pluralism' is not used in the original book nor, therefore, in the English edition, the idea of pluralism plays a central role in the last chapter.

The philosophers and legislators of the late 18th century moved in this direction by granting individuals freedom – theoretically, freedom in all activities compatible with the preservation of society; and liberalistic theory restricts the *laissez-faire* only by demanding the protection of human life and property. But the preservation of society, while not quite as vague an ideal as the interests of a community or the good of mankind or the just, is capable of a multitude of interpretations and implementations which have various degrees of desirability to various individuals; so is even the protection of human life, especially with regard to life's beginning and end; and property or, at any rate, certain kinds of property have been the focus of the greatest controversies and revolutions. Actually, most basic problems along all these lines have so far been solved by one or the other form of monistic (uniform) dictates including, in particular, dictates by majorities.

For many problems, the effectuation of pluralistic solutions is of course severely limited by external circumstances as well as for intrinsic reasons. External problems, which do not concern the solutions as such, arise from historical and geographical facts. Centuries of developments have led to establishments of enormous power which are profoundly interested in preserving the (monistic) *status quo*. The tight geographical and economic interconnections of most parts of today's world and its standardization are further factors strongly favoring monistic regulations. Among the intrinsic difficulties of pluralistic organizations, two are of particular importance. One is the fact that some ideas of social and economic organization as well as some technical projects (the latter epitomized by space exploration) require the cooperation of exceedingly large numbers of participants; and their actualization might be jeopardized by the excessive fragmentation which would result from the superposition of pluralistic regulations (that is, from the organization of groups each of which is cohesive with respect to a great number of questions). The second intrinsic difficulty and perhaps the most serious obstacle to pluralism is the existence of individuals who decide to limit or suppress decisions of others – in a few extreme cases, to exterminate some other groups.

It will be noted however, first of all, that notwithstanding all these difficulties most nations have always included small autonomous or semi-autonomous groups, some of them religious, which regulate various matters according to their own decisions (though many of those groups not only were forced to pay taxes and to do military service for the large surrounding organization but suffered persecution). Especially during the past few decades numerous communes have been founded, for example, in California; and hermits have always existed. At the other end of the spectrum, on the largest possible scale, the capitalistic and the communistic worlds represent two groups arising from unlike economic and social decisions. But in order to *plan* the organization of a large group of people in coexisting cohesive groups without the impediment of an overpowering *status quo* one would have to populate another planet or to reconstruct an isolated part of the earth after some catastrophe or, under the more relaxed conditions of an experiment, to settle an uninhabited island.

Secondly, it may be pointed out that not all aspects of modern life are unfavorable to pluralistic regulations and that some factors may in fact facilitate their effectuation; for example, the possibility of instantaneous communication between any two points on earth, and the existence of computers that permit the sorting and grouping of individuals in large numbers. By these means it has become possible (or certainly will soon be possible) to organize, on the one hand, various groups within the same territory and, on the other hand, groups scattered over altogether separate regions. This technology, available only since the middle of this century, has not yet been exploited for the organization of peacefully coexisting groups with unlike codes.

Thirdly, as to excessive fragmentation through the superposition of too many pluralistic regulations, it is clear, for example, that for the sake of the realization of a grand plan its adherents must make concessions on issues that they consider to be of comparatively minor importance. But equally clear is the possibility of a small number of large groups and a large number of small ones coexisting. Why should there not exist autonomous groups living lives counseled by Tolstoy which, detached from modern civilization, renounce its wasteful affluence and are free of its burdensome obligations and its potential hazards? Such groups might survive modern civilization. Or groups gathered together for purposes of eugenics with aims set by the groups themselves and free to use all the means they see fit to employ? Some such groups might later produce paragons for future generations. Or communities engaging in sexual behavior of their choosing? In the future, one type of such community may be based on a strict monogamy protected by law. Why should not a group carry out, on a voluntary basis and unhampered by outside interference, any social experiments that place no restraints on outsiders? The only rationale for denying existence to such groups would be the unfitness – in some cases the total unfitness – of their codes to serve as rules for everyone. But clearly the negation of *all* and *everyone* is not *nothing* and *no-one*.

Yet social practice often excludes the great middle: *something* and *some people*. Implicitly and perhaps unconsciously, social and legal practitioners usually follow the categorical imperative taking unique regulations for granted. Actually, however, most ethical and social problems have more than one solution, each desired by *some people*; and there are solutions that are incompatible. After emancipation from the spirit of the categorical imperative, what is needed in order to bring that practically "excluded middle" into proper relief is a study which examines cohesive groups associated with unlike decisions, systematically describes the cases where they can be constituted, investigates the relations between such groups, and sets up the conditions under which they can coexist.

Finally, the study just indicated could also contribute to (at least) the alleviation of the principal intrinsic difficulty of pluralistic organizations: decisions of individuals about the decisions of others. Such decisions – as it were, *decisions of second order* – must of course be distinguished from the mere competition of individuals for the same acquisitions or activities. A cohesive group may well include competing individuals, but it cannot include members bent on suppressing the decisions of other members. I believe that there is a wide field for systematic dispassionate investigation (from a logical point of view) of decisions and concessions and of forms of social organization which permit the peaceful coexistence of cohesive groups.

This is what I had in mind when speaking of a logic of ethics.

At the time when this book was written, one could hear the first voices complaining that man's ethical and social development had not been on a par with his scientific and technological progress; and since then, such voices have swelled to a powerful chorus. In 1933, I asked myself what factors in the realm of cognition might be contributing to this lag. Much of what *empirical observation* can supply to influence and motivate decisions had already been revealed in the various branches of biological, psychological, and sociological knowledge. But two other methods that have been immensely successful in natural science were being neglected in dealing with decisions and the groups of their adherents: *analysis* and *experiment*. Even though many reasons make it unlikely that in human affairs these methods will ever yield results comparable in importance to their achievements in natural science, they deserve at least to be tested.

Analysis, systematically applied to relations between groups, may well uncover forms of organization that so far have been overlooked or neglected but which, after being dispassionately considered, may appear highly desirable to some people and influence their decisions. Analysis may, in particular, reveal conditions under which it is possible to establish pluralistic solutions of social problems. (Jurists would undoubtedly find difficulties in non-uniform schemes; but the lag of social behind technological progress will certainly not be diminished without great difficulties being faced and overcome.) Analytic studies may also be justified by the fact that they aim at the formulation of conditions for *the fulfilment of the wishes of as many individuals as possible*; or that they search for a constructive answer to Thoreau's questions, "Is a democracy, such as we know it, the last improvement possible in government? Is it not possible to take a further step towards recognizing and organizing the rights of man?" In addition to this motivation, still another point of view has been taken in this book: pluralistic solutions are looked upon as *schemes that give life to as many ideas as possible.* They enable plans to unfold in unadulterated form, and to materialize through the wholehearted support of adherents without interference from unenthusiastic or obstructing or sabotaging opponents; and, perhaps most important, they make it possible for ideas to *compete* in a manner unobservable and unattainable in monistic solutions, be they dictates by large or small groups or results of compromises. In fact, only pluralistic regulations give ideas a chance comparable to what nature grants organic species. So let some philosophers call the externalization of both value judgments and their underlying ideas superficial! Actually, ideas as such are treated with more sympathy and respect in this book than in the writings of those who propose to subject others to the evaluations and ideas based on their own cognition of absolutes and on their intuition of essences.

An *experiment*, too, has already been alluded to, and a thought-experiment is mentioned in several passages of this book, assuming a somewhat more important role toward the end. One is to imagine a large group of people voluntarily convening, perhaps on an uninhabited island, and deciding on norms, but limiting codes to those who favor them. (A huge stage might simulate the conditions for the initial phase.) A sponsor would have to provide the necessities of life and the means for the continuation of the experiment. In order to be instructive in revealing the consequences of ideas, some experiments would have to extend well beyond one generation. Indeed, patience would be perhaps the chief requirement for participants as well as sponsors. (The element of duration is probably the main reason why even in this age of space exploration experiments in the social field have an utopian character.) To find volunteers to participate in social experiments might well be easier now than it would have been forty years ago since the youth of today are more mature and idealistic.

In limited, if highly important ventures, experimenters might prearrange a partial constitution and confine pluralism to one or two social questions, such as eugenics or the forms of distribution and management of the means of production or the types of compensation for work. On the largest scale, they might start with a legal *tabula rasa*. Whatever their scale and scope, such experiments would embody a synthesis of the main ideas herein propounded: the recantation of the categorical imperative; decisions of individuals as the ultimate basis of rules; the restriction of regulations to their adherents; the formation of cohesive groups; and the establishment of mutually compatible voluntary associations.

Returning to those who complain that ethical and social developments lag behind scientific and technological progress one wonders what they think should happen and how they imagine an advance on the human side might be achieved. One cannot reasonably hope for changes without the labor of changing. And it is hard to envisage progress resulting from ethical studies of the traditional type: from definitions of good and evil or from proofs of the undefinability of these notions; from the intuition of absolute values or from complete skepticism; from metaphysics or from mysticism. Emphasis on pluralistic organizations of society whenever and wherever such regulations are possible is of a different character. Any advance of the social and ethical life, however limited, produced by the effectuation of this suggestion would be based on an analysis of the relationships between (actual and potential) groups and on empirical knowledge, perhaps expanded by experiments, thus on strictly objective thinking with the exclusion of value judgments and metaphysics.

V

Two epistemological topics connected with the content of this book may deserve a brief mention.

First, it should be pointed out – without apology! – that the term 'decision' is not herein defined despite its importance in the study of cohesive groups and voluntary associations or rather *because* of its basic importance. Every communication of ideas clearly includes undefined terms; for the attempt to define all terms would lead to an infinite regress of definitions. The recipient of a communication is free to interpret the undefined terms in his own way. This fact makes most communications equivocal; but it broadens their scope inasmuch as they are compatible with various interpretations (somewhat reminiscent of geometric statements about the undefined terms 'point' and 'line' which are applicable to various objects in various domains: in drafting, to pencil marks and rows of such marks drawn by means of a straightedge; in astronomy, to cross hairs in telescopes and light rays; in kinematics, to particles and uniform motions). Most communications about (undefined) decisions are compatible with the strict behavioristic interpretation which identifies decisions with observable behavior accompanying an action as well as with the introspective view which sees them as the results of weighing conflicting motives. Moreover, the use of the undefined concept is noncommittal toward the question as to whether decisions are "determined" by hereditary and environmental elements, education and preceding experiences or ultimately left to some "free agent."

The second remark concerns the attitude taken in this book toward value judgments. (It will of course be understood that 'value' is here being discussed in its connections with philosophical, ethical and sociological problems and not in its economic context.) Evaluations are painstakingly and, it is hoped, completely avoided.

But is not the exclusion of value judgments itself the result of a decision that is based on an implicit value judgment? And would it not be possible to develop a satisfactory theory of values which would render that exclusion arbitrary? Could one not formulate assumptions including 'value' among their undefined terms and then combine and transform those propositions according to specified rules of logic, thereby obtaining a theory comparable to geometry, which combines and transforms propositions about undefined points and lines? Nowhere does this book deny these possibilities. Of course, such a theory of values would have to be of rigorous logical cohesiveness; and, though nothing resembling the wealth of consequences of geometric postulates would be expected, some interesting conclusions would have to be derived from the assumptions as the *raison d'être* of the theory. But at the time when this book was written, no known theory of value in the ethical-sociological sense met these requirements. In particular, despite its title, Spinoza's *Ethica ordine geometrico demonstrata* is not an example; for neither does that book deal with value judgments nor does it treat its subject matter in a way even remotely meeting modern logical standards.

Incidentally, I felt that if a satisfactory theory of values were to be developed, then attempts to connect it with the social world in the way geometric theories are connected with the physical and astronomical space might well utilize the externalization propounded in this book; for the groups of adherents associated with the value judgments of ethics and sociology certainly are among their strongest links with the social world. But be that as it may, such a theory of values lies outside the scope of this book.

At two places, the book includes allusions to the Vienna Circle, the first at the end of the fourth note in Chapter II. Around 1930, there occurred what I regarded as an inroad of a kind of metaphysics into that group. The development originated with Waismann. For some time he was supported by Carnap and for a longer time, though less emphatically, by Schlick. In the discussions about the foundations of mathematics, Waismann habitually advanced as the *ultima ratio* of his assertions that they evince themselves (*zeigen sich* – a term borrowed from the last part of Wittgenstein's *Tractatus*), while he declared all assertions that went beyond what he said to be meaningless (*sinnlos*).

I repeatedly pointed out that the assertion "a proposition evinces itself" adds to the proposition neither content nor strength while a declaration of its meaninglessness does not detract from it anything that was in it before the declaration. If you consider a proposition to be meaningless, I said to Waismann, then don't assert the proposition or even explicitly state that you will not assert it; but in the absence of a complete set of rules concerning meaning, refrain from calling the proposition meaningless – under the circumstances a dangerous procedure, which risks being called meaningless itself.

I considered Waismann's assertions as kinds of value judgments which, as such, belong in the biographies of their proponents rather than in mathematics and logic. A mathematical theory, I emphasized, consists of nothing but transformations of precisely stated propositions into other propositions according to precisely stated rules, with freedom in the choice of the rules as well as of the propositions.

In 1930, the strong emphasis on that *twofold* freedom was at variance with all the current writings on philosophy of mathematics – not only with the intuitionists' and Russell's ideas, but also with the views held at that time by Carnap, who was still strongly influenced by Wittgenstein and consistently spoke about *the* language and *the* logic in his writings and in discussions. In the Circle, not even Hahn seemed to welcome my deviating ideas. Only Kurt Gödel (who had entered the Circle about the time when I did and thereafter had been a rather untalkative member) assented when in several meetings I presented four papers that I had just published[5]. In those publications I developed the views briefly outlined above, which are essentially what Carnap some years later so aptly called the *Principle of Logical Tolerance*.

The other allusion to the Circle, in Section I of Chapter III, has to do with the Unity of Science movement, which can be traced to Neurath and Carnap and had, from its beginning around 1930, the support of Hahn and Ph. Frank in Prague, while Waismann, Kraft and, in particular, Schlick remained reserved. The movement aimed at the propagation of the idea that the methods of physical science, especially of physics, supply tools for all sound procedures throughout science – science in the broad sense of the German word *Wissenschaft*, which includes humanities. It was a reaction against those schools of social science, then flourishing especially in Germany, which extolled intuition, cognition of essences, and consciousness of absolute values while cavalierly dismissing logical thinking and what is known as scientific method.

Though I fully shared the antipathy of the Circle to all verbiage about essences and the absolute, I was in less than full sympathy with the unity movement. Apart from a general aversion to monistic schemes, I feared

[5] 'Bemerkungen zu Grundlagenfragen. I: Über Verzweigungsmengen', *Jahresber. d. Deutschen Mathematikervereinigung* **37** (1928), 213; 'III. Über Potenzmengen', *ibid.* p. 303. 'Der Intuitionismus', *Blätter f. Deutsche Philosophie* **4** (1930), 311; and the, lecture 'Die neue Logik', in the booklet by H. Mark, H. Thirring, H. Hahn, G. Nöbeling, and K. Menger, *Krise und Neuaufbau in den exakten Wissenschaften. Fünf Wiener Vorträge*, Vienna, 1933. (For an English translation of my lecture, entitled 'The New Logic', cf. *Philosophy of Science* **4** (1937), 299.)

that for the sake of a methodological monism some at least potentially useful procedures might be neglected or even discarded. Moreover, I failed to see – then, as I have ever since – how this unified science and its results could be clearly distinguished from some other, especially from certain artistic, activities and from some works of art. Nor have I ever understood what particularly useful cognitive purpose would be served by such definitory unifications and separations even if they were to be achieved. In this spirit, several passages of the book are devoted to the futility of attempts to define or circumscribe, for example, geometry or ethics.

Despite all these reservations with regard to the unity of science, I felt that the externalization of value judgments supported the general ideas of the Circle – certainly of the Circle as conceived by Schlick.

While writing this book I searched the literature for investigations that might be related to my studies. I found works on ethics and sociology with the most fascinating and instructive material. But most books on *ethical theory* were permeated by value judgments while I eschewed evaluations as a matter of principle and thought that throughout this book I was succeeding in avoiding them completely. In *sociology*, I found only few studies devoted to human groups and relationships; and even these were of a descriptive nature. In fact, my search led me to the work of only the two writers mentioned in Chapter V, the 19th century Austrian G. Ratzenhofer and the contemporary L. von Wiese in Germany. (Through the work of the latter, I later became accquainted with some of the writings of G. Simmel.) But even their studies were essentially empirical rather than applications of exact thinking to empirical material – in today's parlance, to models such as I tried to develop. Ratzenhofer[6], a post-Comte positivist, regarded his sociology as a positive theory of human interrelationships. Von Wiese[7] and his school engaged in systematic studies of human groups and social relationships. They listed a great number of processes and relations arranging and classifying them according to certain principles (some of them not altogether transparent).

[6] *Soziologie. Positive Lehre von den menschlichen Wechselbeziehungen*, Leipzig, 1907.
[7] *Systematic Sociology, on the Basis of the Beziehungslehre and Gebildelehre* (Theory of Relationships and Theory of Structures), New York, 1932. *Sociology: Its History and Main Problems*, Hannover, N.H., 1928.

Von Wiese distinguished processes of the "toward one another" and the "away from one another." The *simple* processes he classified "in increasing order of connection and separation": the former as approximation, adaptation, assimilation and union; the latter as competition, opposition and conflict. Within each of these classes, he listed numerous special processes in alphabetic order. The lists are accompanied by historical, ethnographic and other observations such as "Positive associations are mainly the result of the herd instinct, the urge to build, the sex drive, and curiosity" and by more hazy remarks such as "What lies between birth and death is characterized by a mixture of loneliness and togetherness." Von Wiese's theory of social structures consists of a list of observable organizations and supplementary comments. For example, in discussing triples consisting of a pair approached by a third person he points out (following Simmel) that the two members of the pair almost always react differently to the third person; that the latter sometimes strengthens, sometimes weakens, the relationship between the two; that the triple has the tendency to dissolve, that is, to isolate one member, who then may look for associations outside, so that a double pair may evolve. Other ideas extensively discussed by von Wiese include what he calls the fundamental category of social distance, and the formula $P = A \times S$ purporting to express that a process, P, is the result or product of an attitude, A, and a situation, S. But the exact meaning of the former concept has not become entirely clear to me, and the deductions connected with the formula are certainly logically inadequate. While in the mid 1930's the work of von Wiese and his school did not result in any exact theory of social groups and relationships, their systematic treatment and classification of observable social phenomena compared favorably with the utterly vague discussions of many other sociologists of that period. His methodology owed much to Simmel, whose sociology had dealt with "the web of group affiliations" (*die Kreuzung sozialer Kreise*).

Finally, a remark about the relation of the English edition to the original book. Here and there I have rearranged ideas, especially in integrating into the main text most of the passages which in the original book were set in small print; and where it seemed possible without sacrificing content I have eliminated verbiage, rephrasing and reducing the material – liberties that no translator could have taken. In the third and fourth notes in Chapter IV, I have amended the treatment of duality, supplanted Greek letter notations for groups by more transparent, almost self-explanatory symbols, and replaced equations connecting the fundamental groups by the two diagrams on p. 59sq.

VI

The winter of 1933 and the first half of 1934 during which this book was written belong to the darkest periods in Austrian history. Hitler was consolidating his power in Germany. The near-fascist (though anti-Nazi) Austrian government was acutely threatened by him, his emissaries and the Austrian Nazis. In this situation, the government saw fit to engage in systematic provocations of the most reliable anti-Nazis, the Social Democrats, and, in February 1934, crushed their protests by the use of heavy artillery in the city of Vienna. In July of that year, the Nazis attempted a Putsch and assassinated the chancellor. After quelling this revolt the government, now completely fascist, stayed on for four more years until Hitler's army occupied Austria.

While the University of Vienna was closed for the greater part of 1933/34, the Circle held its informal meetings even during that trying time. But Neurath, who happened to be abroad during the February events never returned to Austria; and at the time of the July revolt, Hahn died. Carnap had gone to Prague in 1931; Feigl had emigrated to the United States and Gödel, more and more estranged from the Circle since the publication of the rather superficial Carnap-Hahn-Neurath manifesto *Wissenschaftliche Weltauffassung* (Scientific View of the World), had stopped attending the meetings even before 1934. Schlick, Waismann (at that time both completely under Wittgenstein's spell) and Kraft were the only members of the original Circle left in Vienna.

While I never mentioned my book in the Circle I discussed parts of it in my Mathematical Colloquium, in which during the spring of 1934 K. Gödel, A. Wald, and F. Alt regularly participated. There, I proved the simple theorem on finite sets[8], mentioned on p. 59. I also showed the manuscript to the economist O. Morgenstern, who even then began studying mathematics and expressed great interest in the book.

[8] Cf. 'Ein Satz über endliche Mengen mit Anwendungen auf die formale Ethik', *Ergebnisse eines Mathematischen Kolloquiums*, Heft 6, Vienna, 1935, p. 23.

After its completion, I gave the manuscript to Schlick and asked him whether the positive ideas in the book were new. To judge from his comments, Schlick read the material carefully. He assured me of the originality of the ideas and only suggested that I might be interested in J. M. Guyau's *Morale sans obligation ni sanction.* I read that booklet with interest indeed, but found, as Schlick had predicted, few points of contact with mine and those just with its critical sections.

When returning my manuscript Schlick offered to publish it in the series of books that he was editing jointly with Ph. Frank and which included his *Problems of Ethics.* I thanked him, but for personal reasons declined the offer. Schlick then said that he would warmly recommend the manuscript to Julius Springer, Vienna, who published the book in the early summer of 1934 under the title *Moral, Wille und Weltgestaltung. Grundlegung zur Logik der Sitten.*

It is impossible to imagine a place and a time less propitious to the publication of a book on ethics than Vienna in the mid 1930's, while during the following years of Nazi domination and world war such a publication would have been of course completely out of the question.

Even before the current revival of interest in the Vienna Circle, reflected in the publication of the Collection devoted to that group by the D. Reidel Publishing Company, I was asked on several occasions to bring out my book in English. (Springer's stock of the original book was destroyed by fire during World War II.) Occupied with questions in various branches of pure mathematics I did not get round to the task despite repeated exhortations from my friend Morgenstern. Suggestions also came from H. Mulder, who has done so very much to keep the interest in the Vienna Circle alive, to broaden and deepen it. I finally devoted the summer and fall of 1973 to the project. Mr. Eric van der Schalie produced a translation of the original which has served me as a basis in preparing the present edition with the most valuable assistance of Ms. Phyllis Kittel.

Chicago, Winter 1973/74

KARL MENGER: PRINCIPAL DATES

1902	born in Vienna
1920–1924	studied at the University of Vienna; Ph.D. in Mathematics
1925–1927	docent at the University of Amsterdam
1927–1936	professor of geometry at the University of Vienna
1930–1931	visiting lecturer at Harvard University and The Rice Institute
1931–1937	founder of the *Ergebnisse eines Mathematischen Kolloquiums*
1932	invited lecturer at the International Congress of Mathematics, Zürich
1936	vice-president of the International Mathematical Congress, Oslo
1937–1946	professor of mathematics at the University of Notre Dame, Indiana
1937–1946	founder of *Reports of a Mathematical Colloquium*, 2nd series, Notre Dame
1946–1971	professor of mathematics, Illinois Institute of Technology, Chicago
1951	visiting lecturer at the Sorbonne, Paris
1961	guest lecturer in several European universities
1964	visiting professor, University of Arizona and Ford Institute, Vienna
1968	visiting professor Middle East Technical University, Ankara
1971	professor emeritus in Chicago

FIELDS OF RESEARCH

I. Theory of Curves and Dimension Theory.
II. Geometry of General Metric Spaces.
III. A General Theory of Length and the Calculus of Variations.
IV. Coordinate-Free Treatment of Curvature.
V. Probabilistic Metric.
VI. Algebra of Geometry. Normed Lattices. Comprehensive Treatment of Diverse Mathematical Structures.
VII. New Foundations for the Bolyai-Lobachevsky Geometry.
VIII. Algebra of Functions. Interconnection of Algebraic Systems.
IX. A New Approach to Calculus. A Theory of the Application of Analysis to Science.
X. Foundations of Mathematics and Logical Tolerance.
XI. Economics: Introduction of Probability into the Theory of Value. The Logic of the Laws of Return.
XII. Ethics.

Collected papers pertaining to the last four sections, supplemented by a complete Bibliography, will appear in a subsequent volume of the Vienna Circle Collection.

PUBLICATIONS IN BOOK FORM

1928: *Dimensionstheorie*, 320 pp., B. G. Teubner, Leipzig.

1932: *Kurventheorie*, 376 pp., B. G. Teubner, Leipzig.

1934: *Moral, Wille und Weltgestaltung*, 144 pp., Springer, Wien.

1940: *Three Lectures on Mathematical Subjects*, The Rice Institute Pamphlets, Vol. 27, No. 1, Houston, Texas.

1952: *You will like Geometry*, 26 pp., Museum of Science and Industry, Chicago.

1954: *Géométrie Générale*, 80 pp., Gauthier-Villars, Paris.

1955: *Calculus. A Modern Approach*, XVIII+354 pp., Ginn & Co., Boston.

1970: (With L. M. Blumenthal) *Studies in Geometry*. Parts 2 and 4, 216 pp., W. H. Freeman & Co., San Francisco.

VIENNA CIRCLE COLLECTION

1. OTTO NEURATH, *Empiricism and Sociology*. Edited by Marie Neurath and Robert S. Cohen. With a Selection of Biographical and Autobiographical Sketches. Translations by Paul Foulkes and Marie Neurath. 1973, xvi+473 pp., with illustrations. ISBN 90 277 0258 6 (cloth), ISBN 90 277 0259 4 (paper).
2. JOSEF SCHÄCHTER, *Prolegomena to a Critical Grammar*. With a Foreword by J. F. Staal and the Introduction to the original German edition by M. Schlick. Translated by Paul Foulkes. 1973, xxi+161 pp. ISBN 90 277 0296 9 (cloth), ISBN 90 277 0301 9 (paper).
3. ERNST MACH, *Knowledge and Error. Sketches on the Psychology of Enquiry*. Translated by Paul Foulkes. ISBN 90 277 0281 0 (cloth), ISBN 90 277 0282 9 (paper).
5. LUDWIG BOLTZMANN, *Theoretical Physics and Philosophical Problems. Selected Writings*. With a Foreword by S. R. de Groot. Edited by Brian McGuinness. Translated by Paul Foulkes. ISBN 90 277 0249 7 (cloth), ISBN 90 277 0250 0 (paper).
6. KARL MENGER, *Morality, Decision, and Social Organization. Toward a Logic of Ethics*. With a Postscript to the English edition by the Author. Based on a translation by E. van der Schalie. ISBN 90 277 0318 3 (cloth), ISBN 90 277 0319 1 (paper).